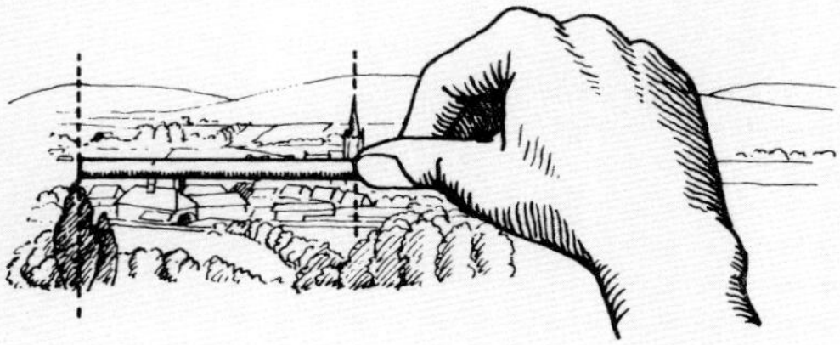

The True Line

The Landscape Diagrams of Geoffrey Hutchings

Colin Sackett | 2006

Foreword

I first saw these printed diagrams and drawings over thirty years ago, and the particular care and certainty they convey has remained with me since.

Hutchings published just a handful of books, all addressing the search for geographical and topographical truths, and for the ways of recording and depicting these truths precisely and economically by the handwritten word and line.

In addition to his contribution to the development of the teaching of field studies in Britain in the late 1940s, with its emphasis on the direct observation and interpretation of landscape, he achieved a masterly ability to 'read' and transcribe a place in a graphic composition—be it a sketch-map or a plan, a tabular profile or a section, or an annotated panoramic drawing. In all of these compositions he integrated line and text in a perfect balance of brevity and detail.

Published Sources

—Printed in *An Introduction to Geographical Landscape Drawing*, 1955

Title page: Measurement with a pencil held at arm's length. (Fig.8)

—Printed in *Great Britain, Essays in Regional Geography*, 1928

1–2: The South-East: geology. (Fig.3)

3–6: A tabular summary of the geology of the South-East. (Fig.4)

7–8: Block-diagram of the Wealden drainage. (Fig.6)

9–10: A dry-valley profile. (Fig.7)

11, 13: The South-East: a diagram showing the relationship of certain aspects of the region to one another and to the geological outcrops. (Fig.12)

—Printed in *An Introduction to Regional Surveying*, 1930

12, 14: A specimen transect chart, Limpsfield, Surrey. (Fig.22)

—Printed in *London's Countryside*, 1957

15: The broad elements in the structure of South-east England. (Fig.1)

16, 18: Section across the London Basin and the Weald. (Fig.3)

17: Generalized geological map of the country between the Chilterns and the Central Weald. (Fig.2)

19, 21: Profiles of the Hythe Beds escarpment as seen in the view westward from Leith Hill. (Fig.9)

20: The Mole basin. (Fig.10)

22: Successive courses of the Thames during and after the Ice Age. (Fig.15)

23, 25: Rivers of the western and central Weald. (Fig.14)

24, 26: Long profiles showing a possible correlation of the Mole terraces with those of the Thames. (Fig.12)

27, 29: Transect of the North Downs (Ashtead–Headley–Betchworth, Surrey). (Fig.20)

28: Parish boundaries on the North Downs in Surrey. (Fig.17)

30: Sketch-map of Headley and Box Hill, Surrey. (Fig.19)

31: Diagram illustrating features of the Headley valley. (Fig.21)

32, 34: Section of the North Downs and Greensand upland (Ranmore–Leith Hill, Surrey). (Fig.26)

33: Geological map of the country between Dorking and Leith Hill, Surrey (after H A Hayward). (Fig.25)

35, 37: Section of the North Downs (Albury–East Clandon, Surrey). (Fig.27)

36, 38: Geological map of the country south of the Medway gap. (Fig.29)

Landscape Diagrams

The reproductions have been originated from the pages of the printed books. In just a few cases the letterpress caption has been digitally deleted in order to make blank the space around the figure.

1

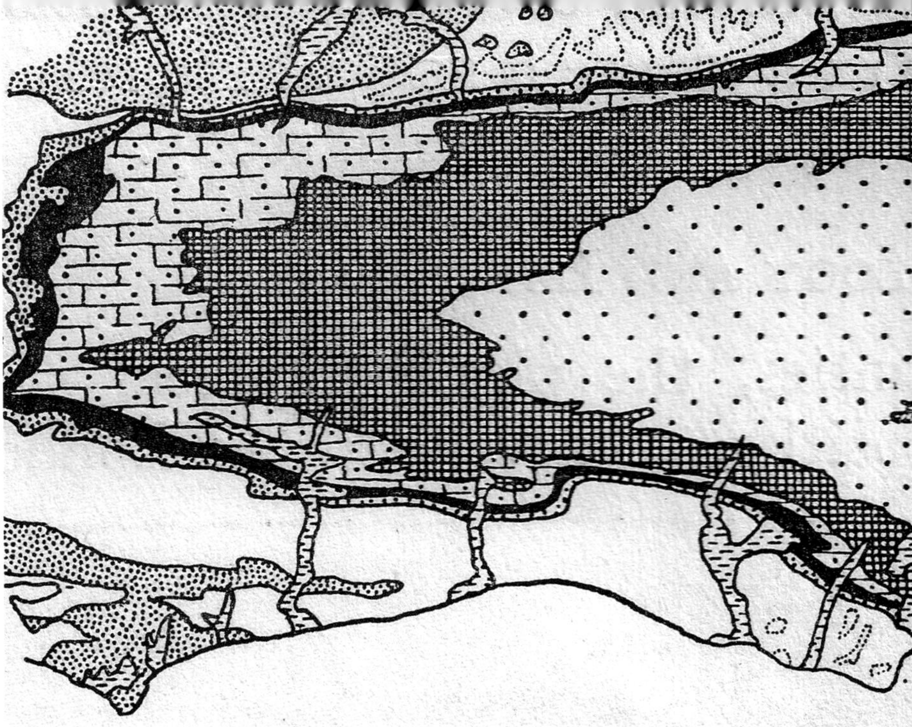

2

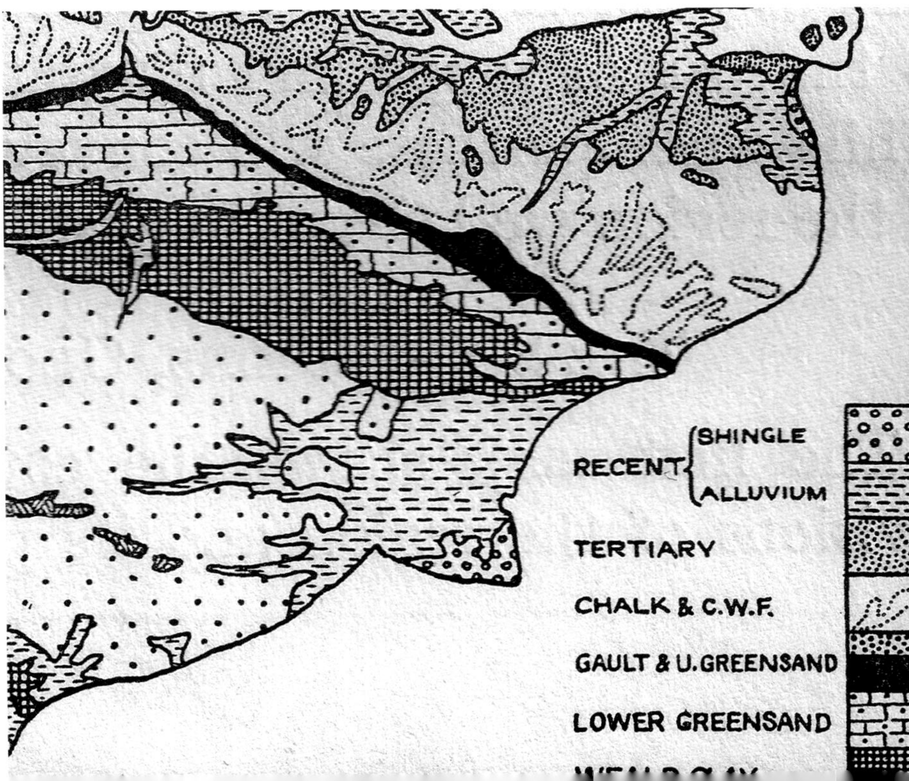
RECENT
SHINGLE
ALLUVIUM
TERTIARY
CHALK & C.W.F.
GAULT & U.GREENSAND
LOWER GREENSAND

3

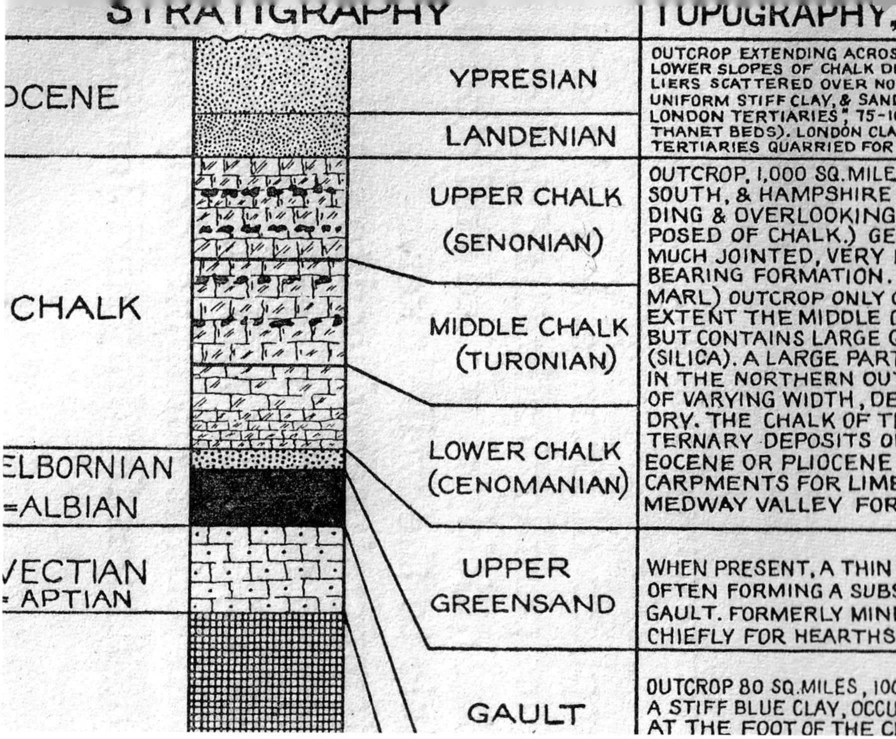
STRATIGRAPHY
TOPOGRAPHY,
OCENE
YPRESIAN
LANDENIAN
CHALK
UPPER CHALK
(SENONIAN)
MIDDLE CHALK
(TURONIAN)
LOWER CHALK
(CENOMANIAN)
ELBORNIAN
=ALBIAN
VECTIAN
= APTIAN
UPPER
GREENSAND
GAULT
WHEN PRESENT, A THIN
OFTEN FORMING A SUBS
GAULT. FORMERLY MINE
CHIEFLY FOR HEARTHS

4

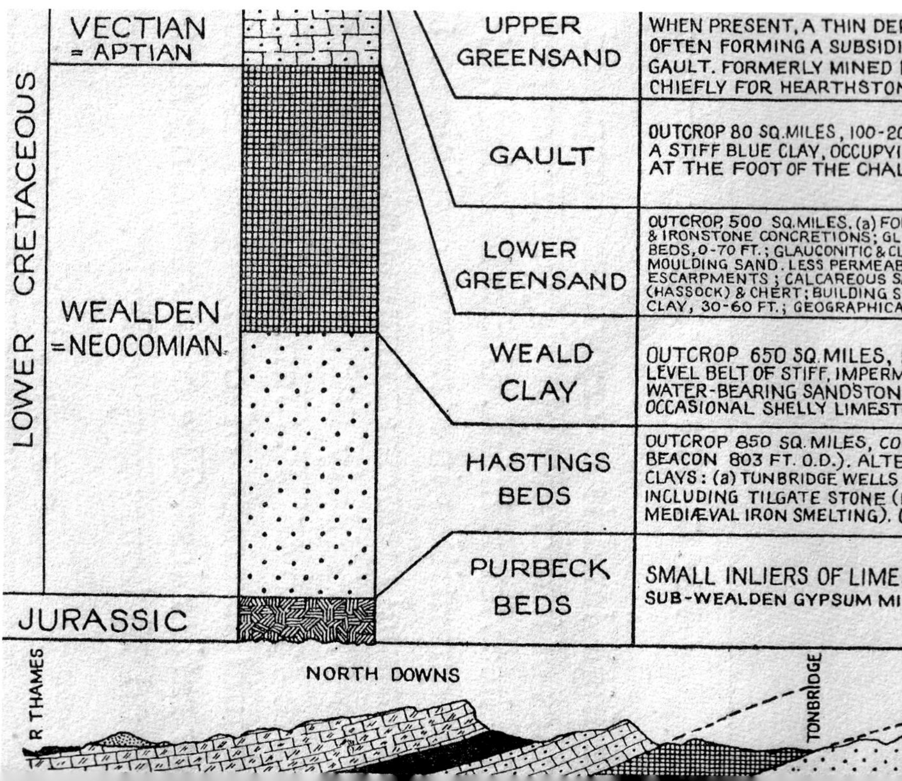
LOWER CRETACEOUS
VECTIAN
= APTIAN
WEALDEN
=NEOCOMIAN
JURASSIC
UPPER
GREENSAND
GAULT
LOWER
GREENSAND
WEALD
CLAY
HASTINGS
BEDS
PURBECK
BEDS
SMALL INLIERS OF LIME
SUB-WEALDEN GYPSUM MI
R THAMES
NORTH DOWNS
TONBRIDGE

ODUCTS

SOME PARTICULAR SECTIONS

5

ING
OUT-
OF
OWR
ND
LON.

TH,
JN-
M-
NE,
ER-
ALK
ESS
NE,
NT
LY
EAU
IOW
A-
BY
S-
HE
DS

ONE
K &
NOW

ER)
LE
TERY

...IN LOCALLY BY THIN DEPOSITS OF DRIFT. THESE INCLUDE: ...Y-WITH-FLINTS" ON THE PLATEAU OF THE NORTH DOWNS, ...CENE) IN THE RIVER VALLEYS & ALONG THE ESTUARINE BELT.

1. THANET TO ROMNEY MARSH.

2. SHEPPEY TO MARDEN

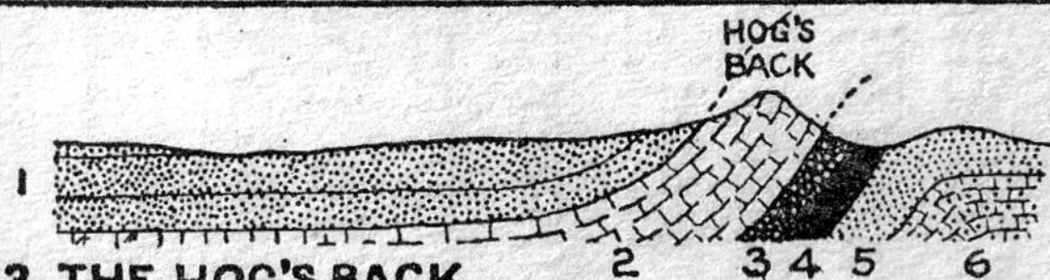

3. THE HOG'S BACK

1. FOLKESTONE BEDS. 2. SANDGATE BEDS. 3. HYTHE BEDS. 4. ATHERFIELD CLAY.

DIAGRAMMATIC SECTION SHOWING LOWER GREENSAND SUBDIVISIONS

6

AREOUS SANDSTONE
WEEN THE CHALK &
DING STONE, BUT NOW

LLY MUCH THICKER)
NE BEDS THE VALE
CKS & ROUGH POTTERY

CK; FERRUGINOUS SAND
RMEABLE. (b) SANDGATE
Y FOR FULLERS EARTH &
FT.; HIGH LAND & STEEP
GLAUCONITIC SANDS
ABLE. (d) ATHERFIELD
NG WEALD CLAY.

BROAD, LOW-LYING,
DINATE RIDGES OF
ROOFING SLABS) AND
ICKS, TILES, POTTERY.

WEALD (CROWBORO'
ANDS & IMPERMEABLE
URST CLAY, 100-180 FT
TONE (USED IN THE
T. — CHALYBEATE SPRINGS

ENTRAL WEALD

THE OUTCROPS OF ALL FORMATIONS ARE OVERLAIN LOCALLY BY T... SANDS & BRICKEARTHS OF PLIOCENE AGE & "CLAY-WITH-FLINTS"... ALLUVIAL BRICKEARTHS & GRAVELS (PLEISTOCENE) IN THE RIV...

1. FOLKESTONE BEDS. 2. SANDGATE BEDS. 3. HYTHE BEDS 4. ATHER CLAY.

DIAGRAMMATIC SECTION SHOWING LOWER GREENSAND SUBDIVIS

1. ASHDOWN SAND. 2. WADHURST CLAY. 3. TUNBRIDGE WELLS SAND

4. LAMBERHURST. DETAILED SECTION SHOWING SUBDIVISIONS OF HASTINGS BEDS

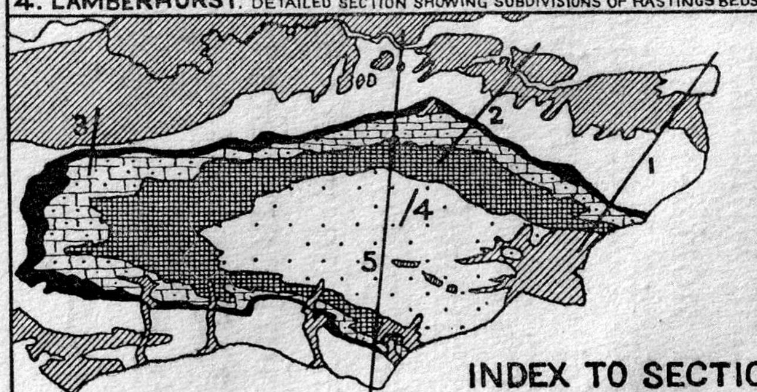

INDEX TO SECTIC

SOUTH DOWNS

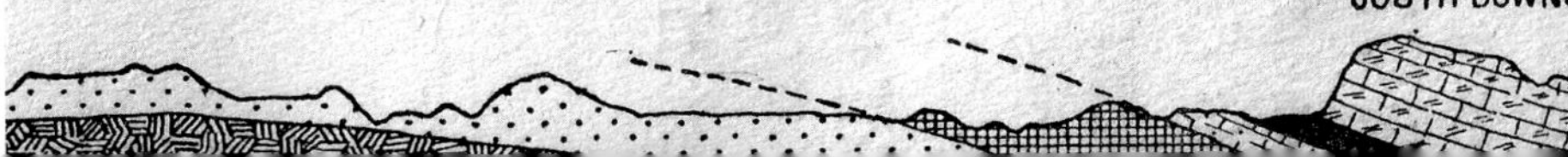

7

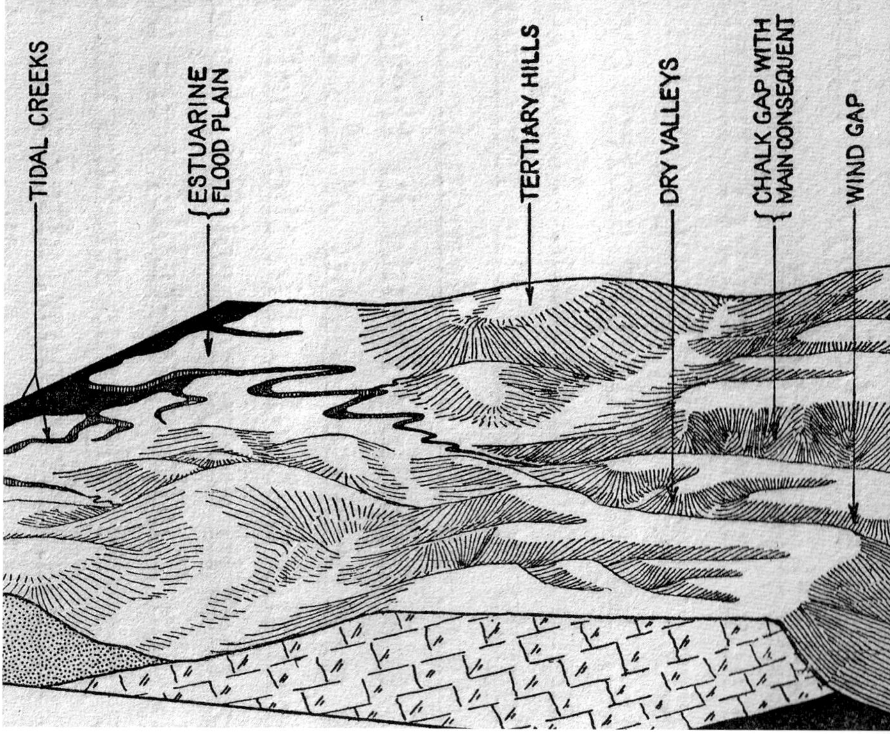
TIDAL CREEKS
ESTUARINE FLOOD PLAIN
TERTIARY HILLS
DRY VALLEYS
CHALK GAP WITH MAIN CONSEQUENT
WIND GAP

8

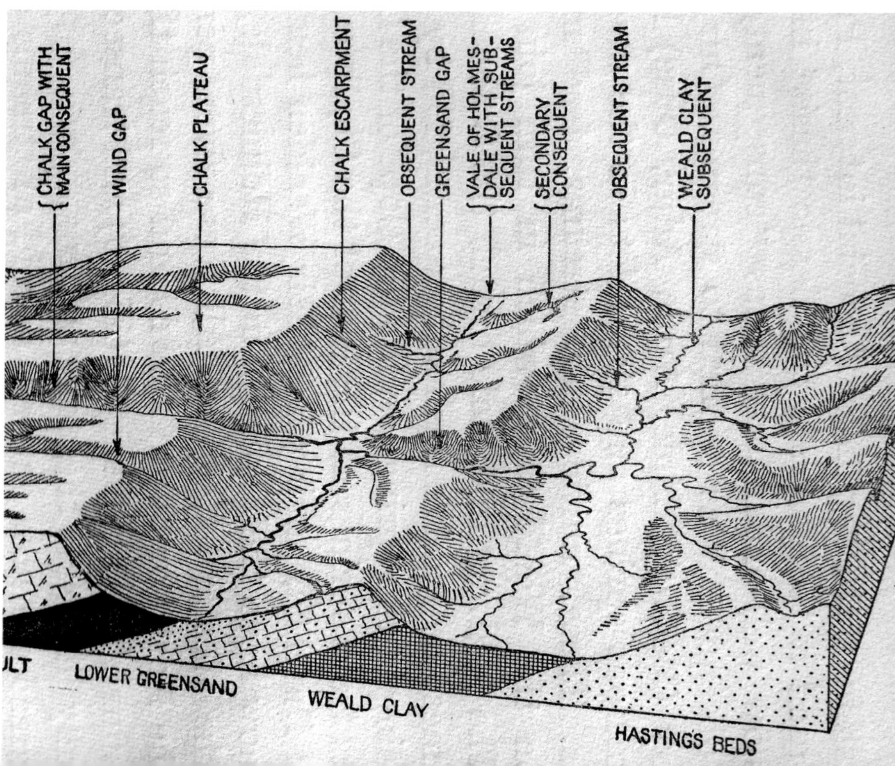
CHALK GAP WITH MAIN CONSEQUENT
WIND GAP
CHALK PLATEAU
CHALK ESCARPMENT
OBSEQUENT STREAM
GREENSAND GAP
VALE OF HOLMES-DALE WITH SUB-SEQUENT STREAMS
SECONDARY CONSEQUENT
OBSEQUENT STREAM
WEALD CLAY SUBSEQUENT
LOWER GREENSAND
WEALD CLAY
HASTINGS BEDS

9

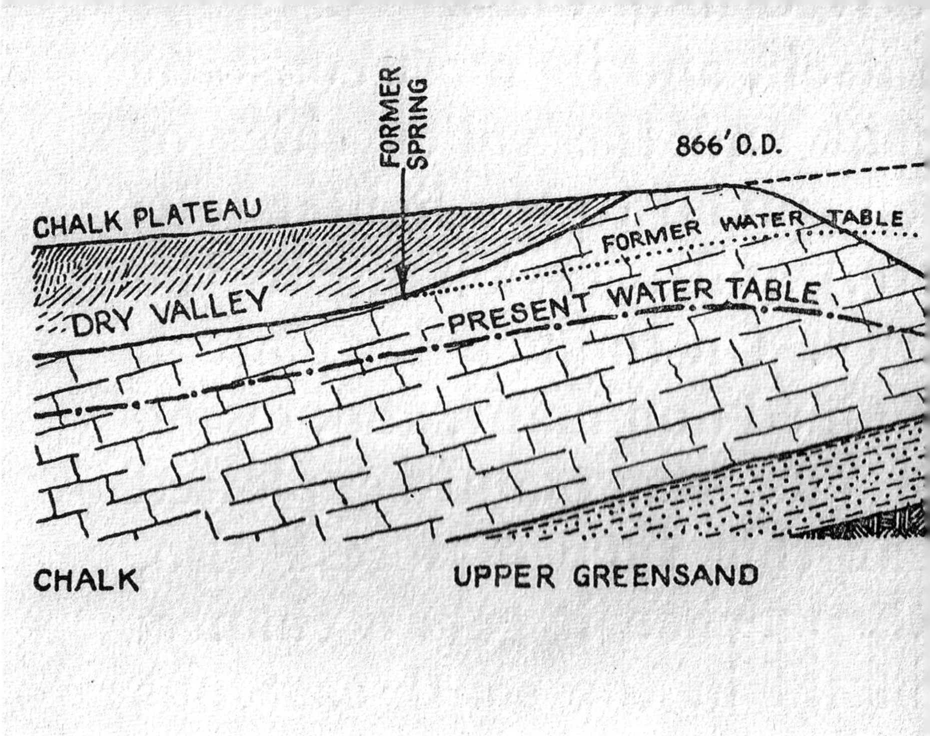
FORMER SPRING
866′ O.D.
CHALK PLATEAU
FORMER WATER TABLE
DRY VALLEY
PRESENT WATER TABLE
CHALK
UPPER GREENSAND

10

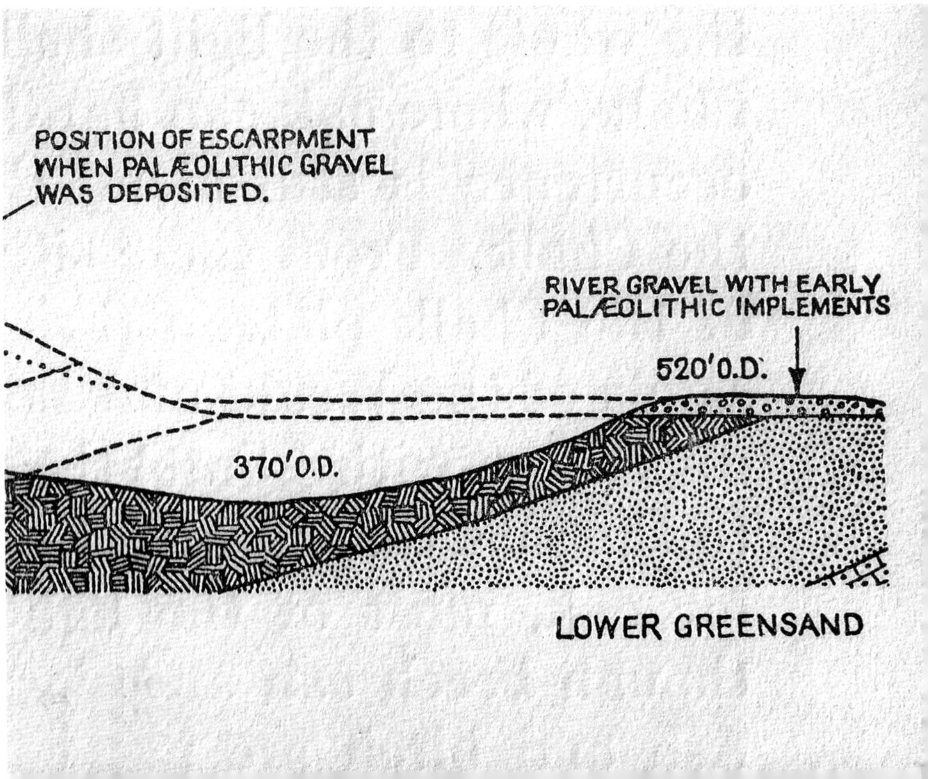
POSITION OF ESCARPMENT WHEN PALÆOLITHIC GRAVEL WAS DEPOSITED.
RIVER GRAVEL WITH EARLY PALÆOLITHIC IMPLEMENTS
520′ O.D.
370′ O.D.
LOWER GREENSAND

19

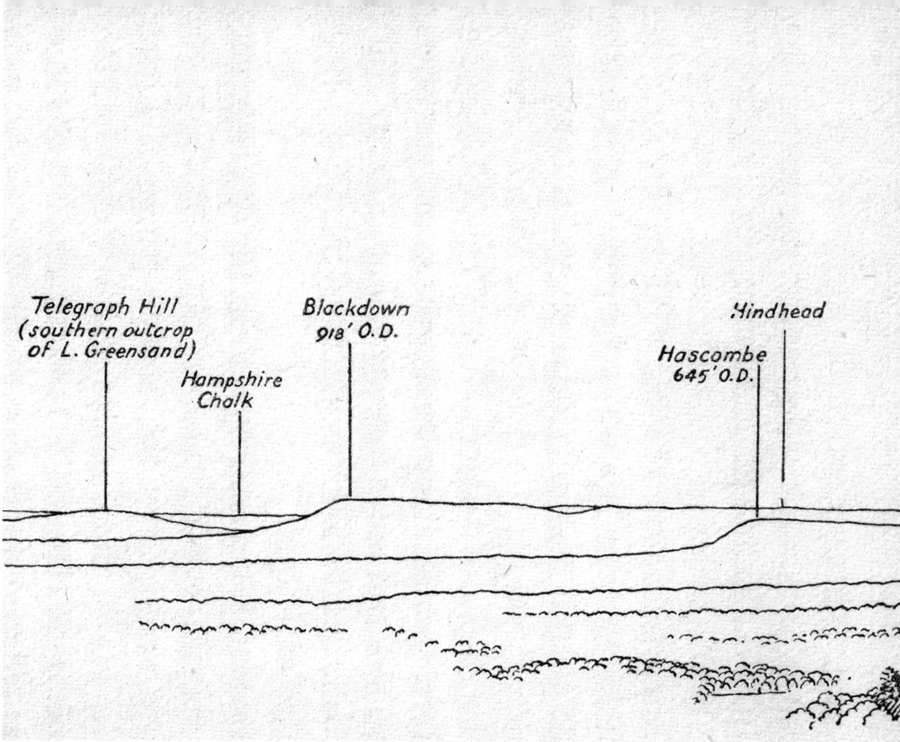

Telegraph Hill
(southern outcrop
of L. Greensand)
Hampshire
Chalk
Blackdown
918' O.D.
Hascombe
645' O.D.
Hindhead

20

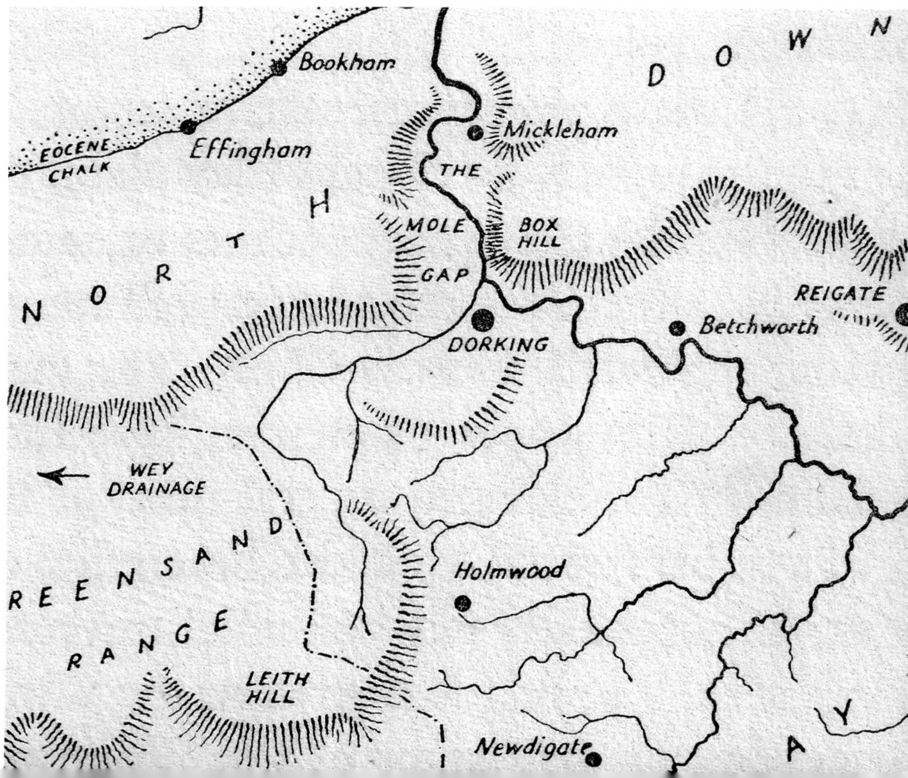

Bookham
EOCENE
CHALK
Effingham
Mickleham
THE
MOLE
GAP
BOX
HILL
D O W N
N O R T H
REIGATE
DORKING
Betchworth
WEY
DRAINAGE
REENSAND
RANGE
LEITH
HILL
Holmwood
Newdigate
A Y

21

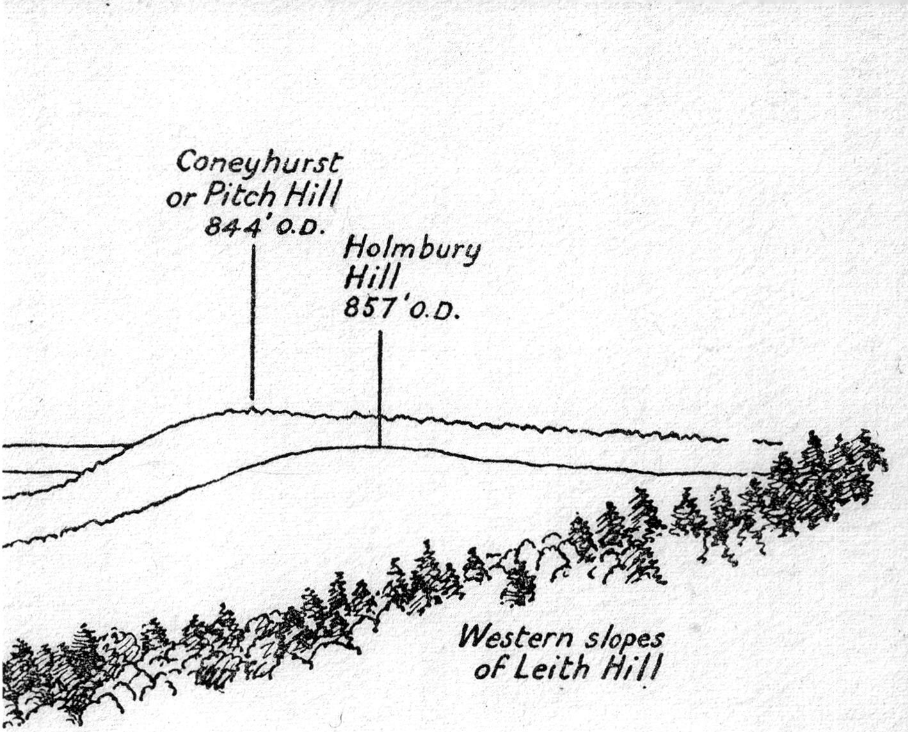
Coneyhurst
or Pitch Hill
844' O.D.
Holmbury
Hill
857' O.D.
Western slopes
of Leith Hill

22

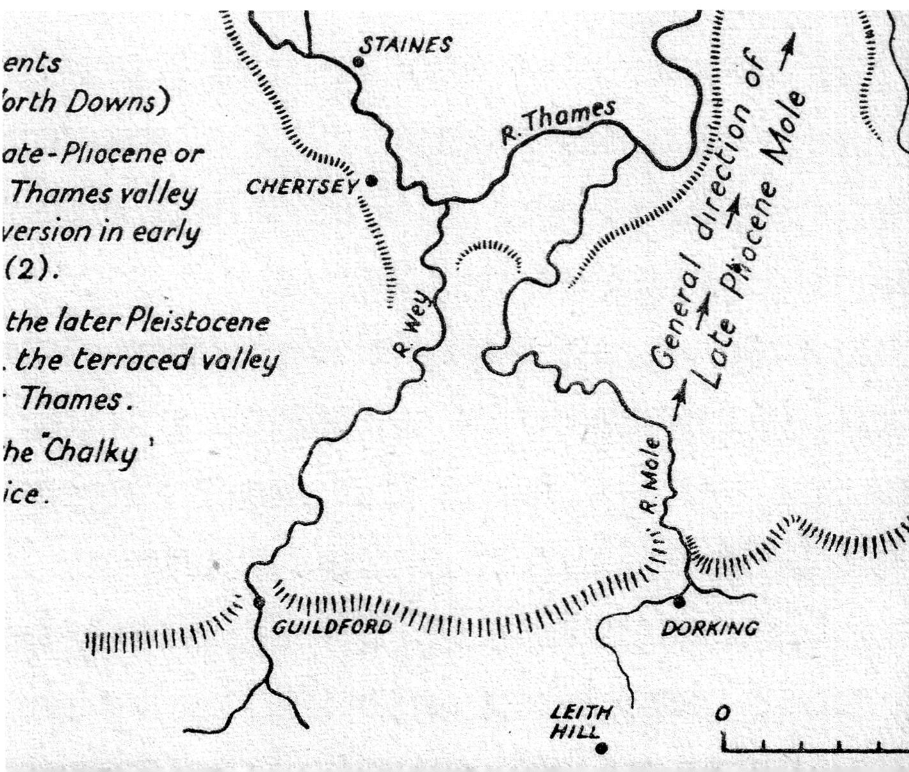
ents
forth Downs)
ate-Pliocene or
Thames valley
version in early
(2).
the later Pleistocene
. the terraced valley
Thames.
he "Chalky"
ice.
STAINES
R. Thames
CHERTSEY
R. Wey
General direction of
Late Pliocene Mole
R. Mole
GUILDFORD
DORKING
LEITH
HILL
0

23

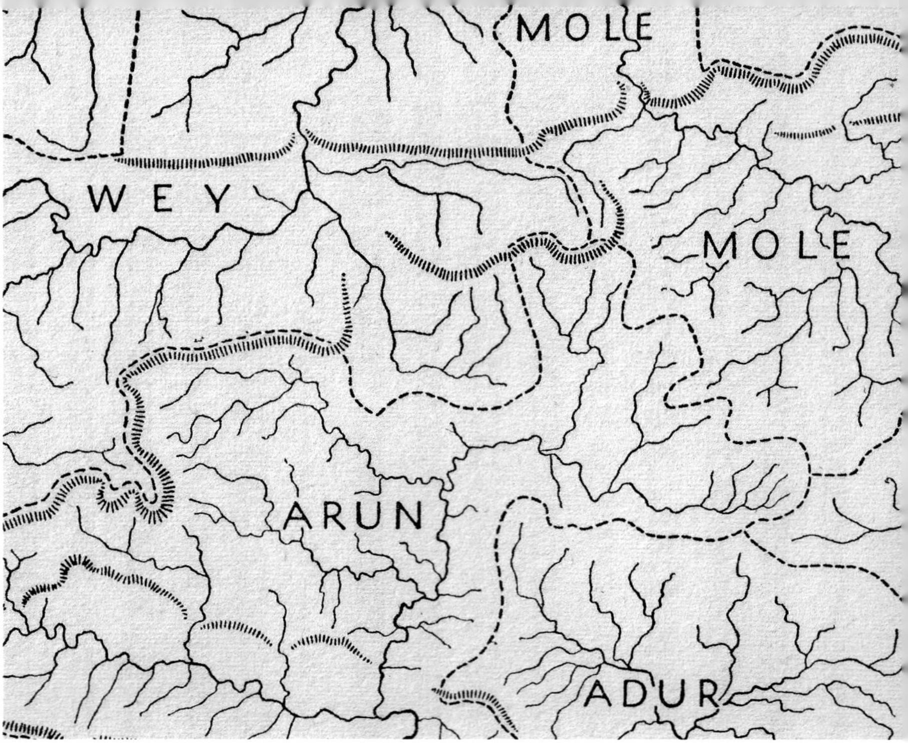
MOLE
WEY
MOLE
ARUN
ADUR

24

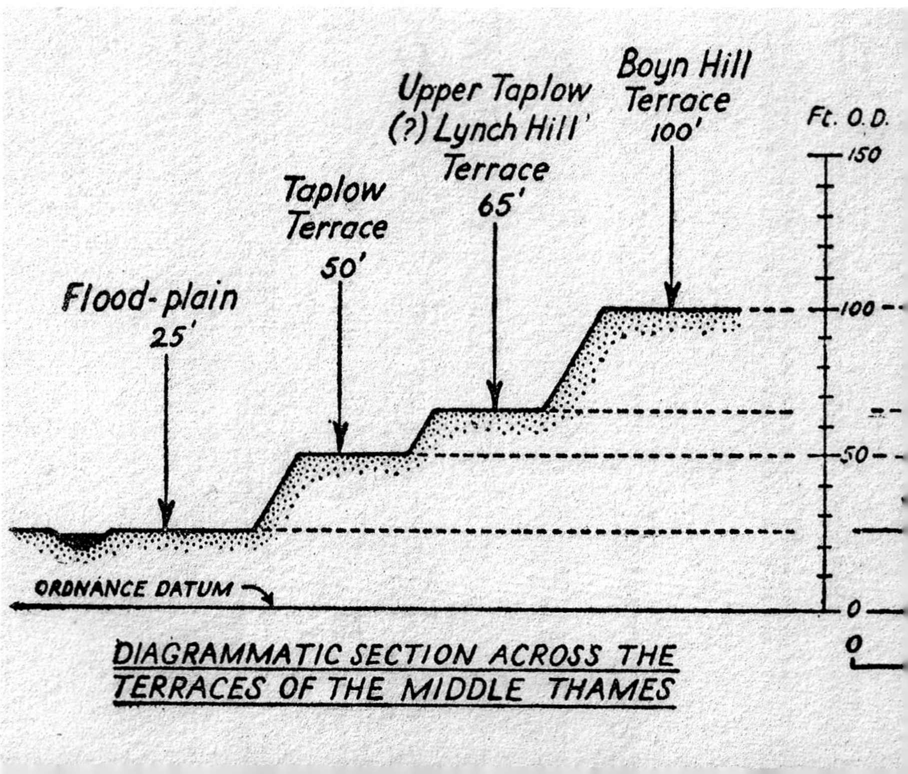
Boyn Hill
Terrace
100'
Upper Taplow
(?) Lynch Hill'
Terrace
65'
Taplow
Terrace
50'
Flood-plain
25'
Ft. O.D.
150
100
50
0
ORDNANCE DATUM
DIAGRAMMATIC SECTION ACROSS THE
TERRACES OF THE MIDDLE THAMES

25

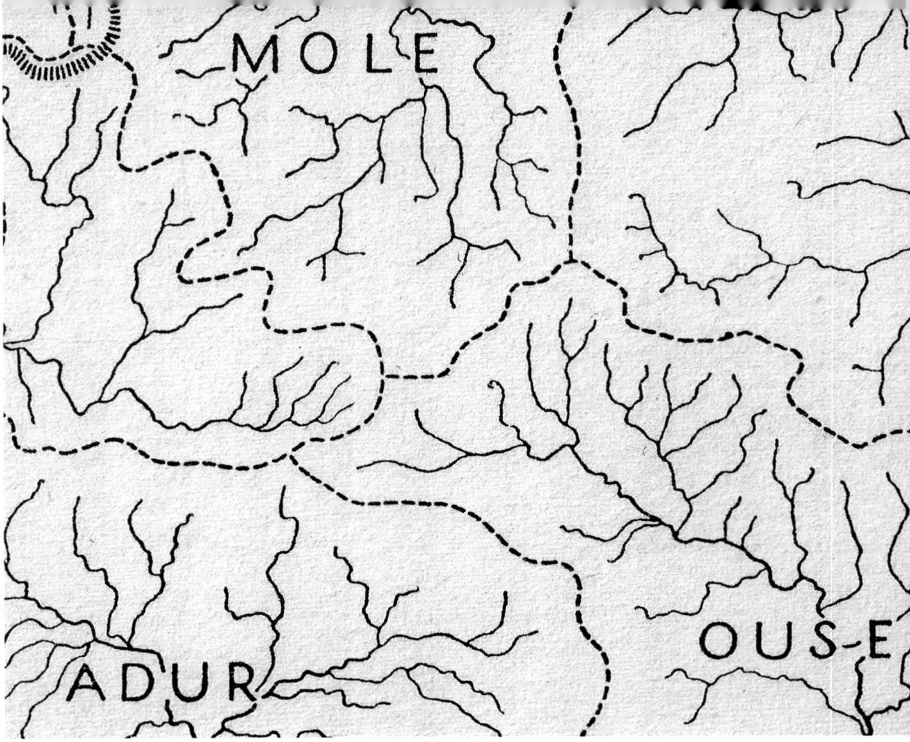

26

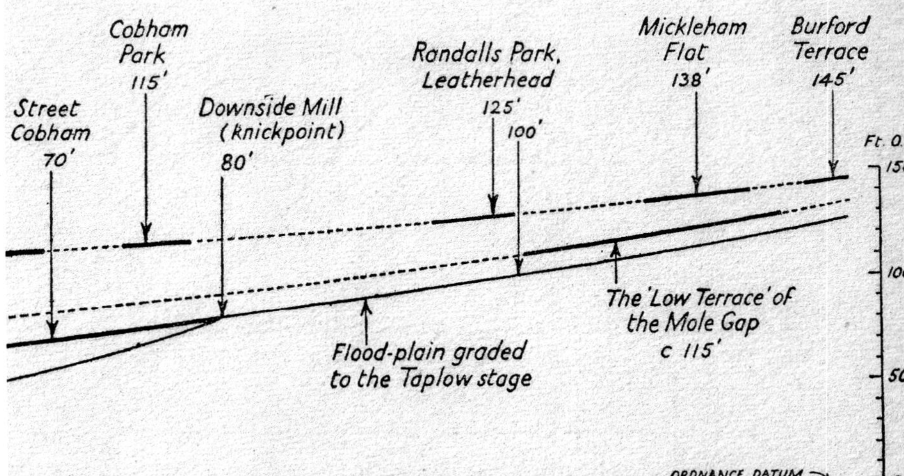

...e purposes of this graph horizontal distances are measured along an assumed centre-line of ...alley, taking into account the major incised meanders but ignoring the course of the river ...the sinuous form of the margins of the terraces and flood-plain. G.E.H.

27

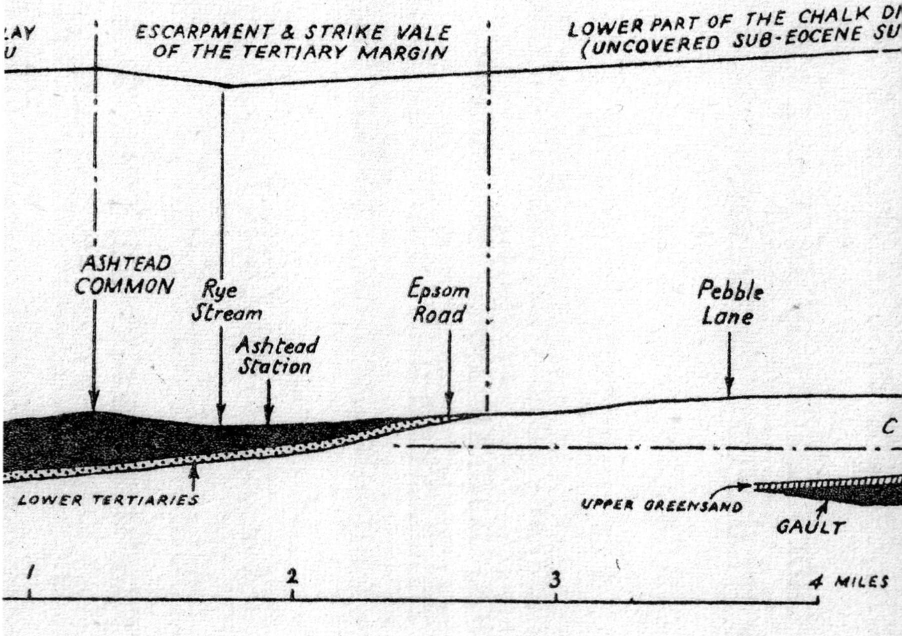
ESCARPMENT & STRIKE VALE
OF THE TERTIARY MARGIN
ASHTEAD
COMMON
Rye
Stream
Ashtead
Station
Epsom
Road
Pebble
Lane
C
LOWER TERTIARIES
UPPER GREENSAND
GAULT
1
2
3
4 MILES

28

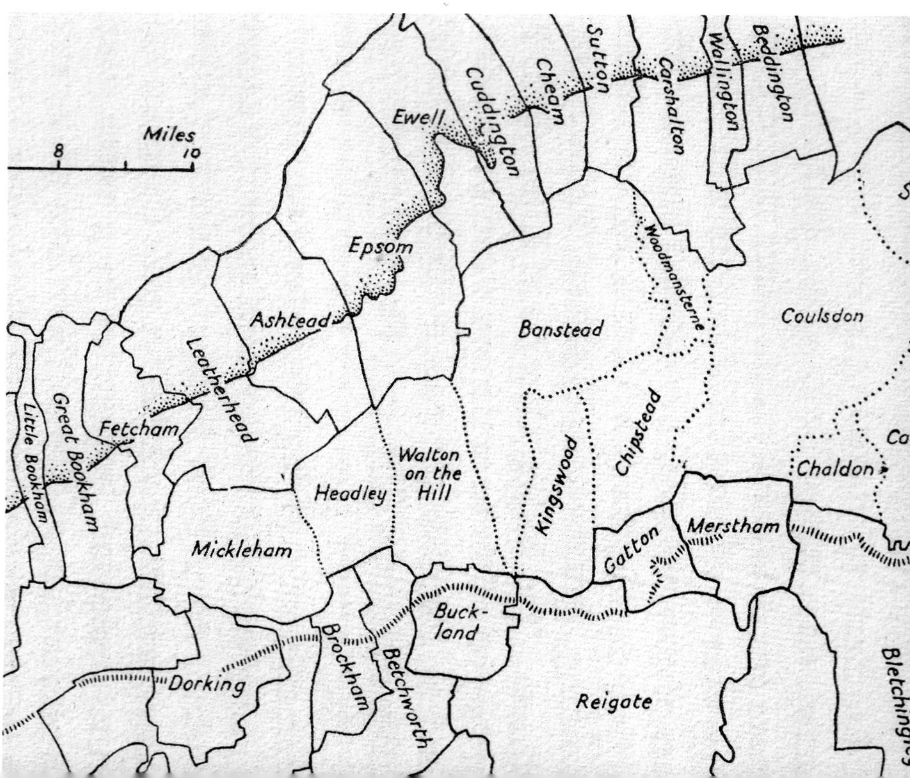
Miles
8
10
Sutton
Cheam
Cuddington
Ewell
Carshalton
Wallington
Beddington
Epsom
Woodmansterne
Ashtead
Banstead
Coulsdon
Leatherhead
Fetcham
Great Bookham
Little Bookham
Walton
on the
Hill
Kingswood
Chipstead
Chaldon
Headley
Mickleham
Gatton
Merstham
Buck-
land
Brockham
Betchworth
Dorking
Reigate

29

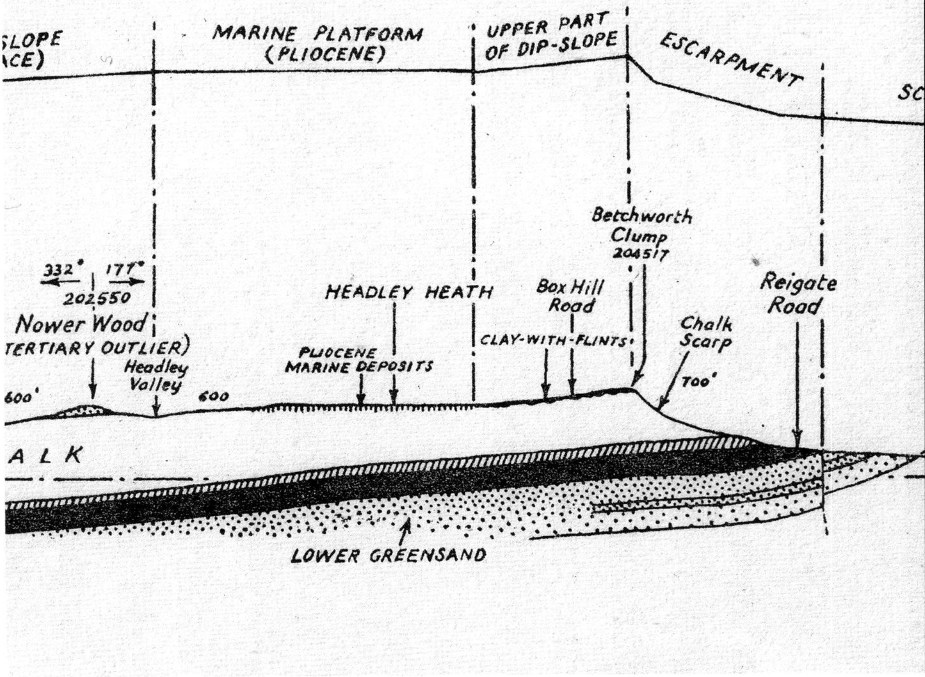
SLOPE
ACE)
MARINE PLATFORM
(PLIOCENE)
UPPER PART
OF DIP-SLOPE
ESCARPMENT
Betchworth
Clump
204517
332°
177°
202550
HEADLEY HEATH
Box Hill
Road
Reigate
Road
Nower Wood
TERTIARY OUTLIER)
CLAY-WITH-FLINTS
Chalk
Scarp
Headley
Valley
PLIOCENE
MARINE DEPOSITS
700'
600'
600
ALK
LOWER GREENSAND

30

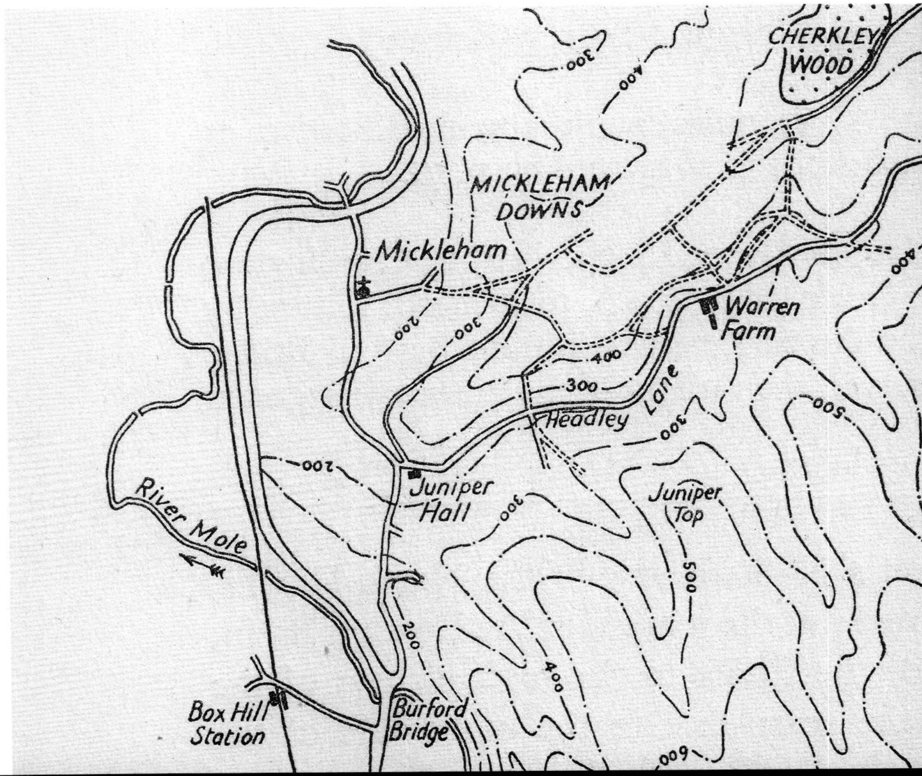
CHERKLEY
WOOD
MICKLEHAM
DOWNS
Mickleham
Warren
Farm
Headley
Lane
Juniper
Hall
Juniper
Top
River Mole
Box Hill
Station
Burford
Bridge

31

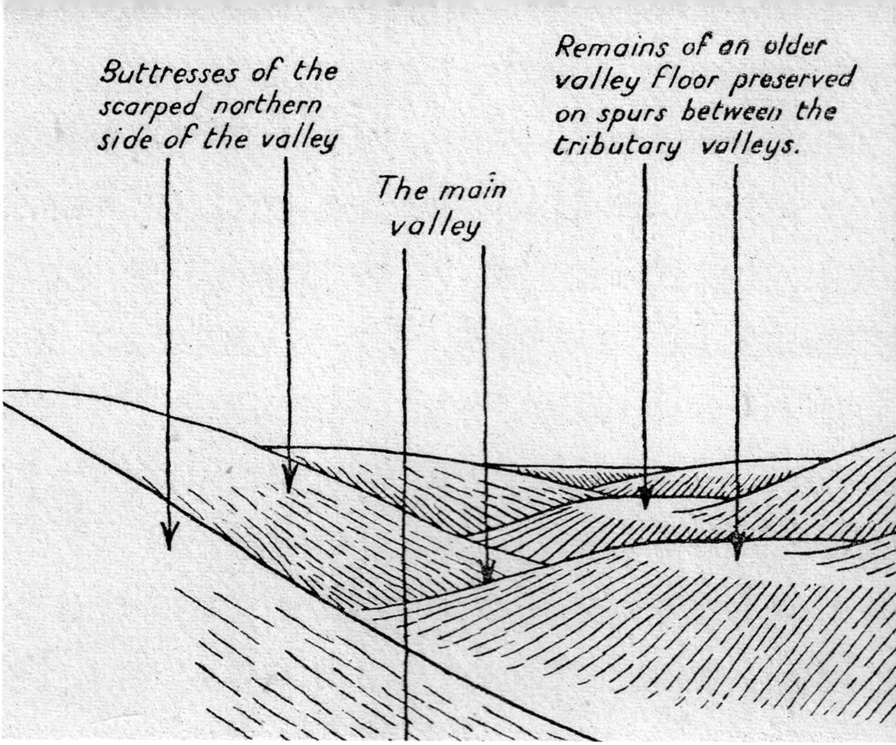

Buttresses of the scarped northern side of the valley
Remains of an older valley floor preserved on spurs between the tributary valleys.
The main valley

32

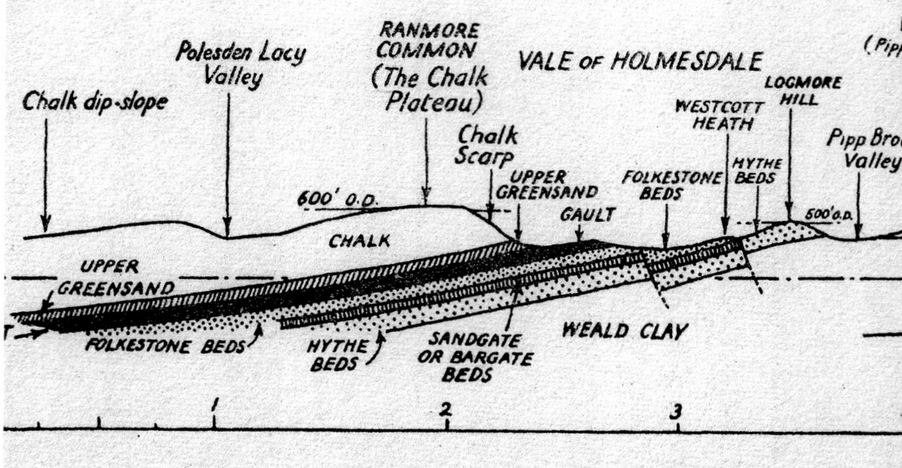

RANMORE COMMON (The Chalk Plateau)
Polesden Lacy Valley
VALE OF HOLMESDALE
Chalk dip-slope
LOGMORE HILL
WESTCOTT HEATH
Chalk Scarp
HYTHE BEDS
UPPER GREENSAND
FOLKESTONE BEDS
600' O.D.
GAULT
500' O.D.
CHALK
UPPER GREENSAND
WEALD CLAY
FOLKESTONE BEDS
HYTHE BEDS
SANDGATE OR BARGATE BEDS
1
2
3

33

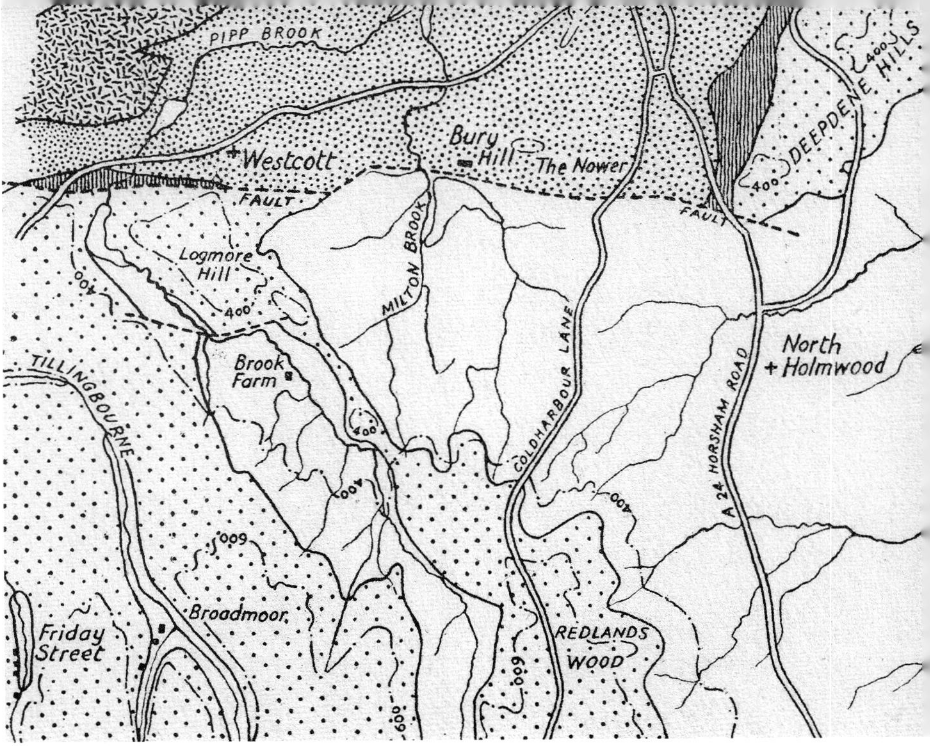
PIPP BROOK
Westcott
Bury Hill
The Nower
DEEPDENE HILLS
400
FAULT
FAULT
Logmore Hill
400
MILTON BROOK
COLDHARBOUR LANE
A 24 HORSHAM ROAD
North
Holmwood
TILLINGBOURNE
Brook Farm
Broadmoor
Friday Street
REDLANDS WOOD
600

34

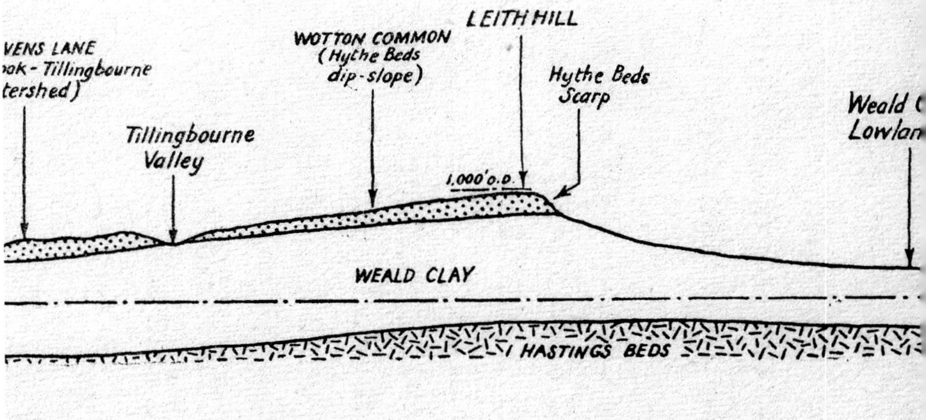
LEITH HILL
WOTTON COMMON
(Hythe Beds
dip-slope)
Hythe Beds
Scarp
Tillingbourne
Valley
1,000' O.D.
WEALD CLAY
HASTINGS BEDS

35

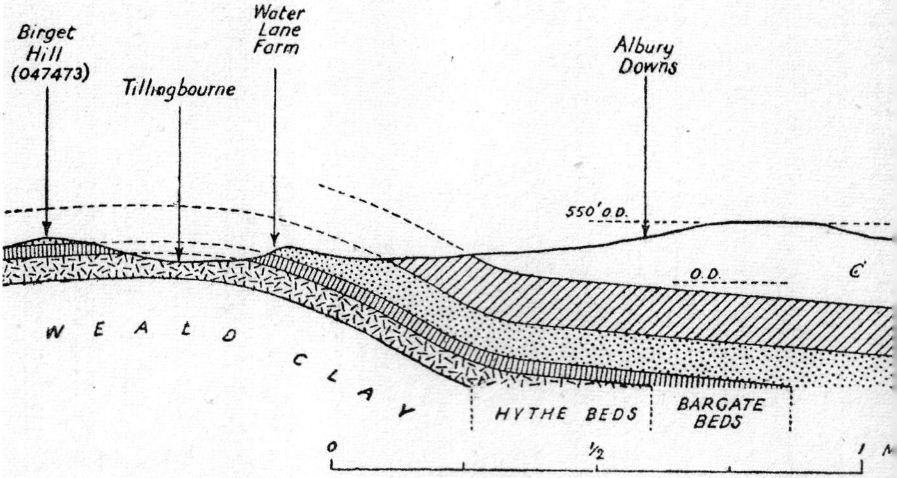

Birget Hill (047473)
Tillingbourne
Water Lane Farm
Albury Downs
550' O.D.
O.D.
W E A L D C L A Y
HYTHE BEDS
BARGATE BEDS
0
½
1

36

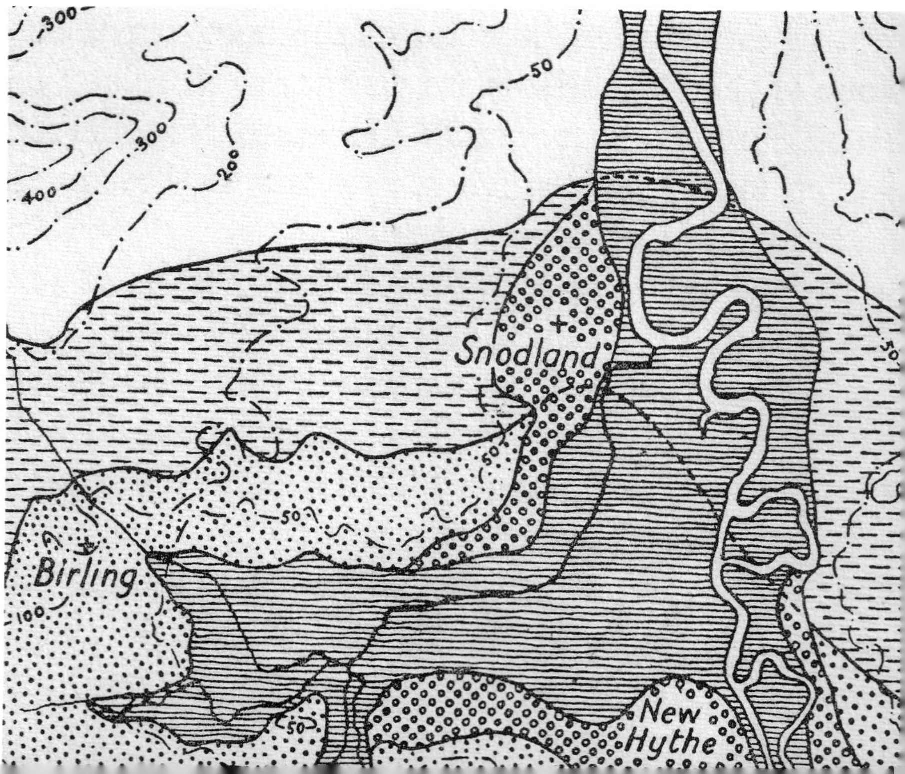

300
400
200
50
Snodland
Birling
100
New Hythe

37

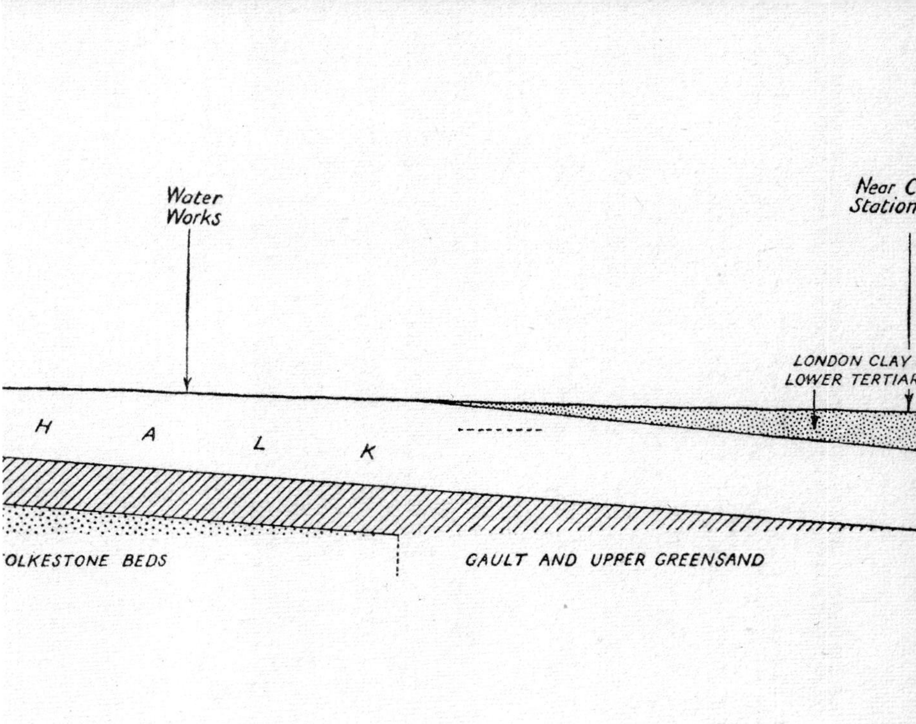
Water Works
Near C
Station
LONDON CLAY
LOWER TERTIAR
H A L K
OLKESTONE BEDS
GAULT AND UPPER GREENSAND

38

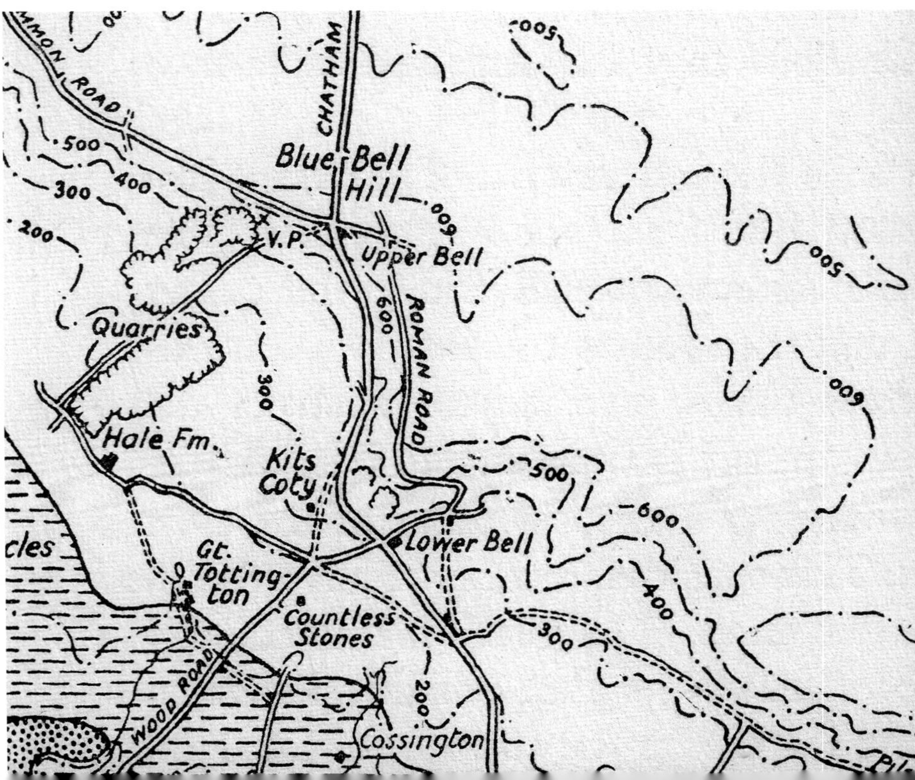
ROAD
CHATHAM
Blue Bell Hill
V.P.
Upper Bell
Quarries
ROMAN ROAD
Hale Fm.
Kits Coty
Lower Bell
cles
Gt. Totting-ton
Countless Stones
WOOD ROAD
Cossington
500
400
300
200
600

39

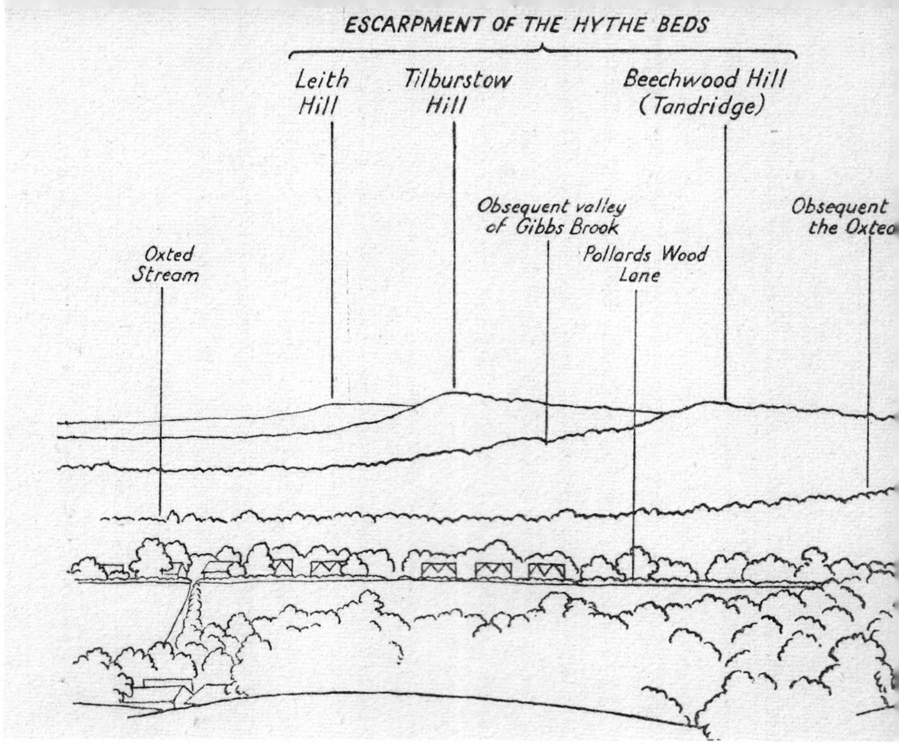
ESCARPMENT OF THE HYTHE BEDS
Leith Hill
Tilburstow Hill
Beechwood Hill (Tandridge)
Obsequent valley of Gibbs Brook
Obsequent
the Oxte
Oxted Stream
Pollards Wood Lane

40

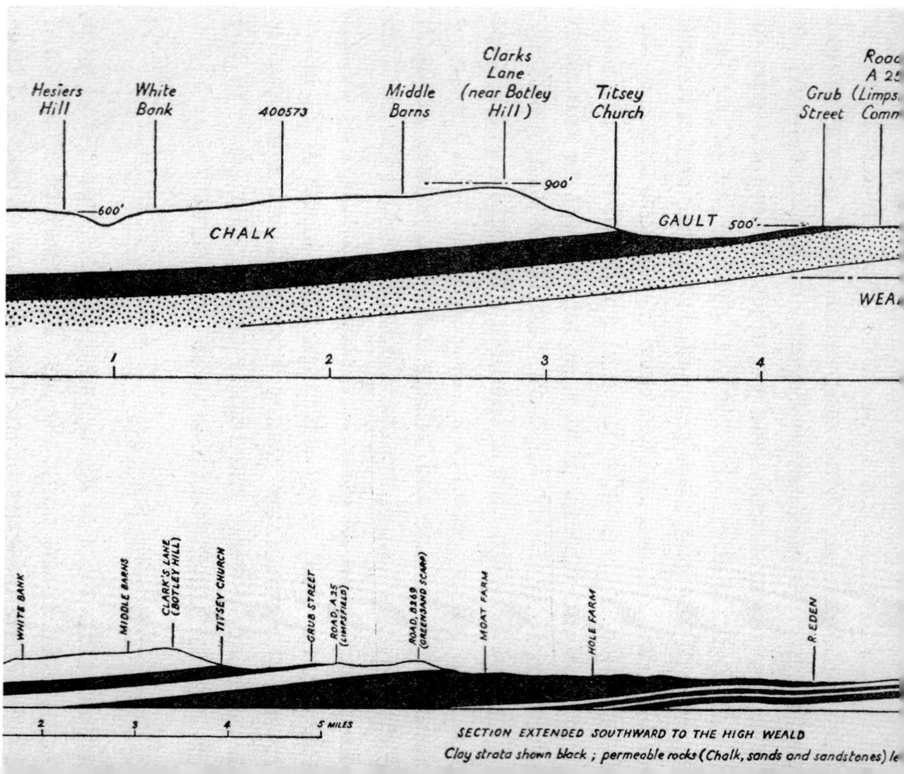
Hesiers Hill
White Bank
400573
Middle Barns
Clarks Lane (near Botley Hill)
Titsey Church
Grub Street
Road A 2
(Limps
Comm
900'
600'
CHALK
GAULT
500'
WEA
1
2
3
4
WHITE BANK
MIDDLE BARNS
CLARK'S LANE (BOTLEY HILL)
TITSEY CHURCH
GRUB STREET
ROAD, A 25 (LIMPSFIELD)
ROAD, B 269 (GREENSAND SCARP)
MOAT FARM
HOLE FARM
R. EDEN
2
3
4
5 MILES
SECTION EXTENDED SOUTHWARD TO THE HIGH WEALD
Clay strata shown black ; permeable rocks (Chalk, sands and sandstones) le

41

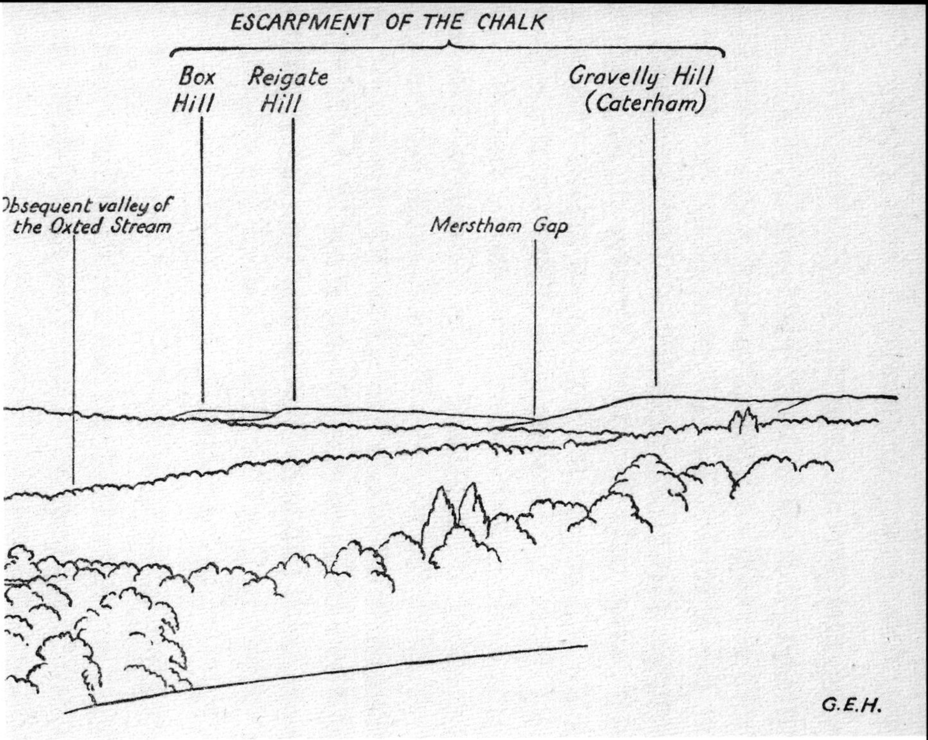
ESCARPMENT OF THE CHALK
Box Hill
Reigate Hill
Gravelly Hill (Caterham)
Obsequent valley of the Oxted Stream
Merstham Gap
G.E.H.

42

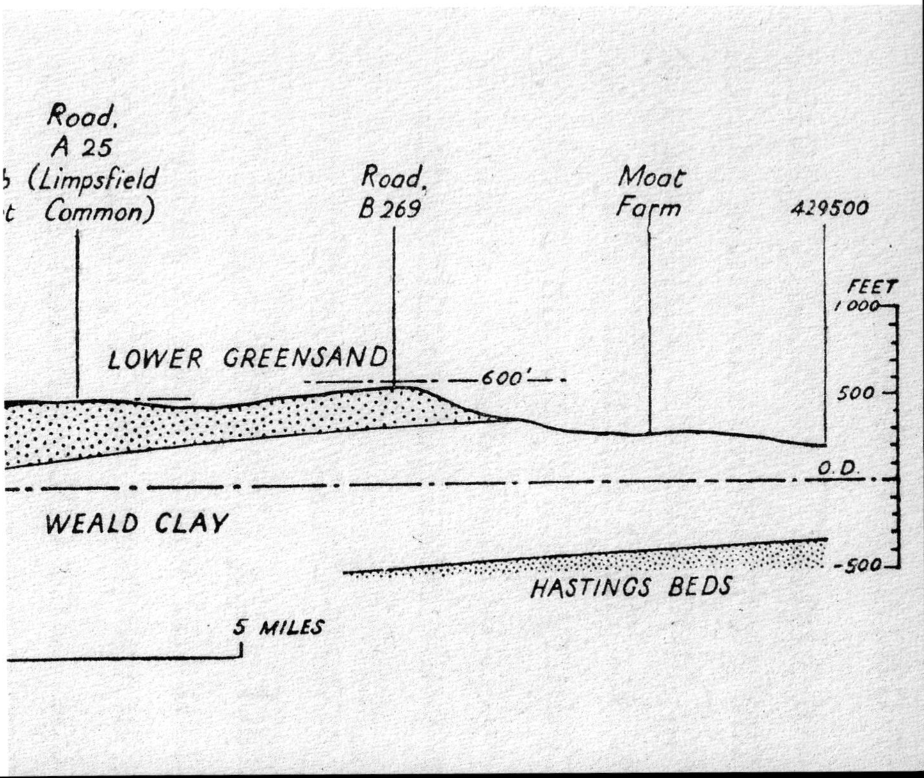
Road. A 25
(Limpsfield Common)
Road. B 269
Moat Farm
429500
FEET
1 000
500
O.D.
-500
LOWER GREENSAND
600'
WEALD CLAY
HASTINGS BEDS
5 MILES

43 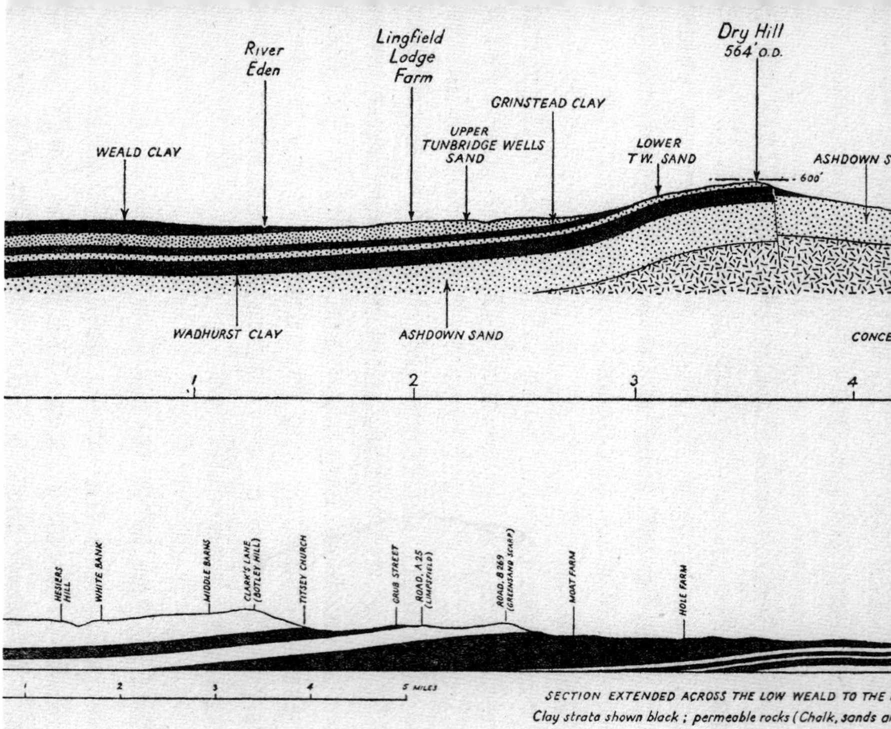
River Eden
Lingfield Lodge Farm
Dry Hill 564' O.D.
GRINSTEAD CLAY
UPPER TUNBRIDGE WELLS SAND
LOWER T.W. SAND
WEALD CLAY
600'
WADHURST CLAY
ASHDOWN SAND
1
2
3
4
HESIERS HILL
WHITE BANK
MIDDLE BARNS
CLARK'S LANE (BOTLEY HILL)
TITSEY CHURCH
GRUB STREET
ROAD, A25 (LIMPSFIELD)
ROAD, B269 (GREENSAND SCARP)
MOAT FARM
HOLE FARM
1
2
3
4
5 MILES
SECTION EXTENDED ACROSS THE LOW WEALD TO THE
Clay strata shown black ; permeable rocks (Chalk, sands

44 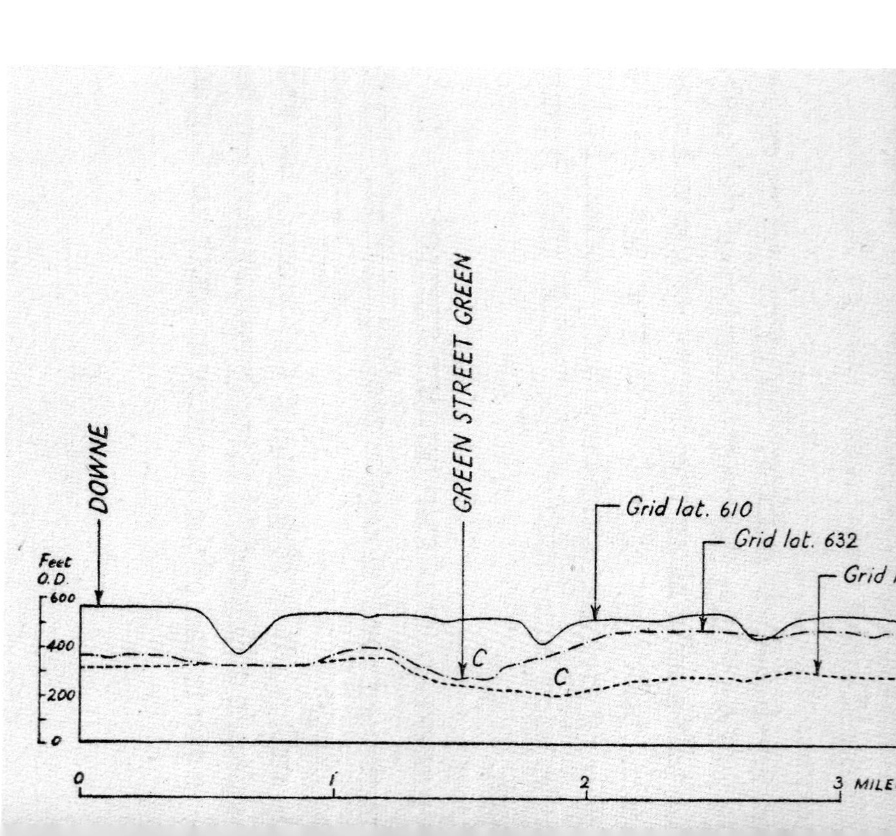
DOWNE
GREEN STREET GREEN
Grid lat. 610
Grid lat. 632
Feet O.D.
600
400
200
0
C
C
0
1
2
3

45

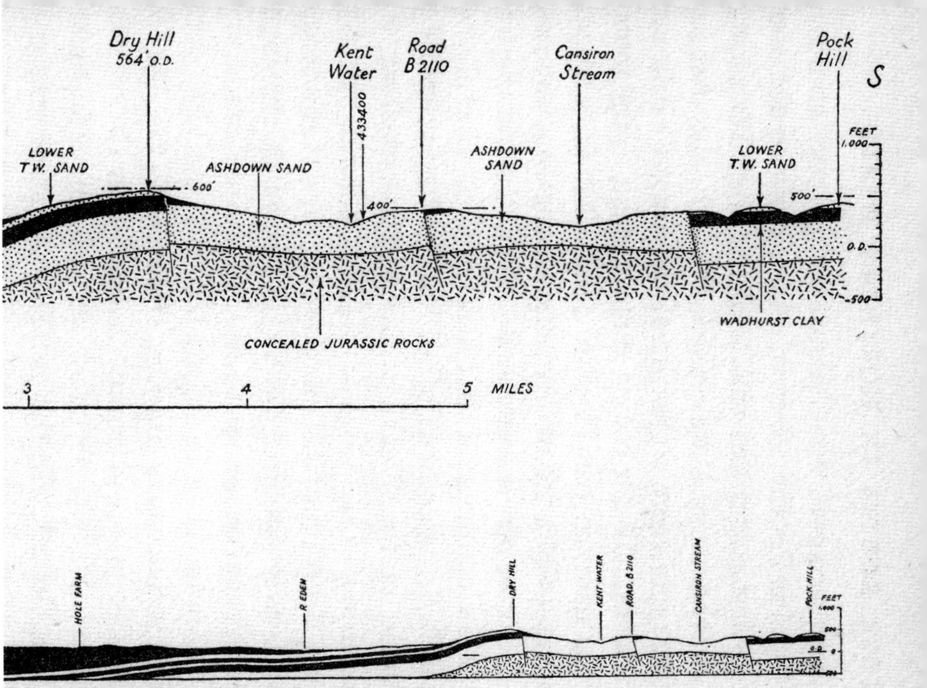

EXTENDED ACROSS THE LOW WEALD TO THE NORTHERN ESCARPMENTS
shown black ; permeable rocks (Chalk, sands and sandstones) left white. G.E.H.

46

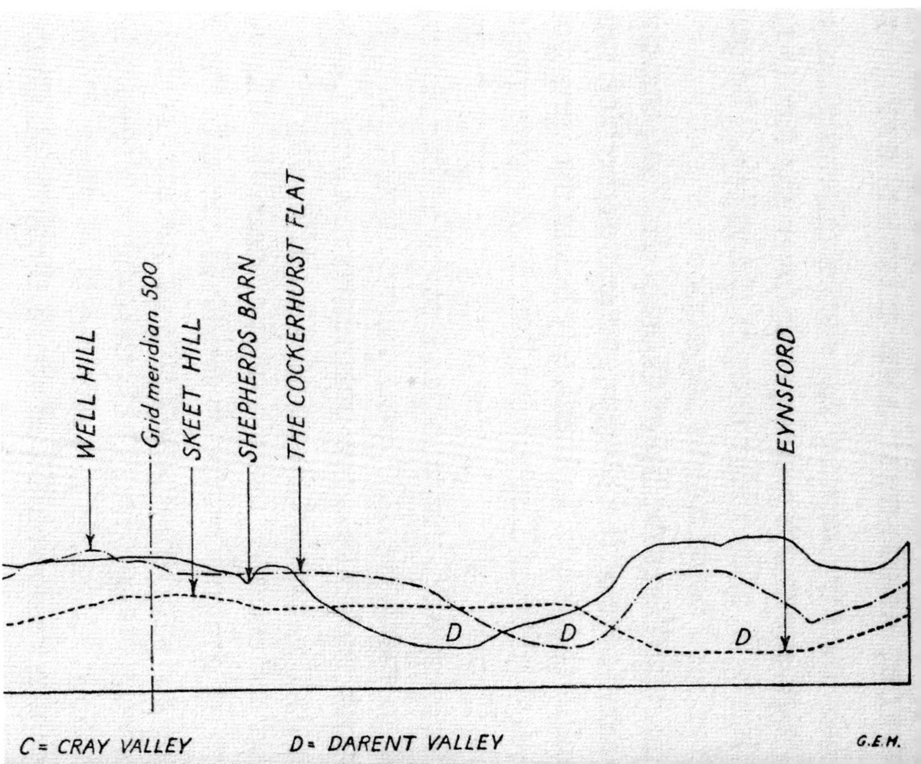

47

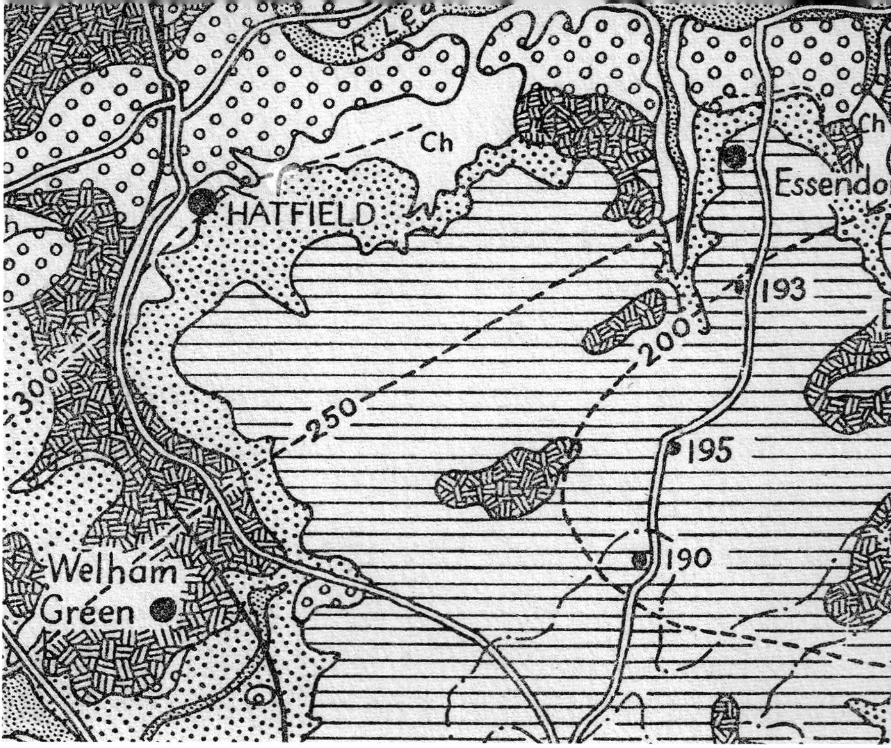

R Lea
Ch
HATFIELD
Essendo
193
200
250
300
195
190
Welham
Green

48

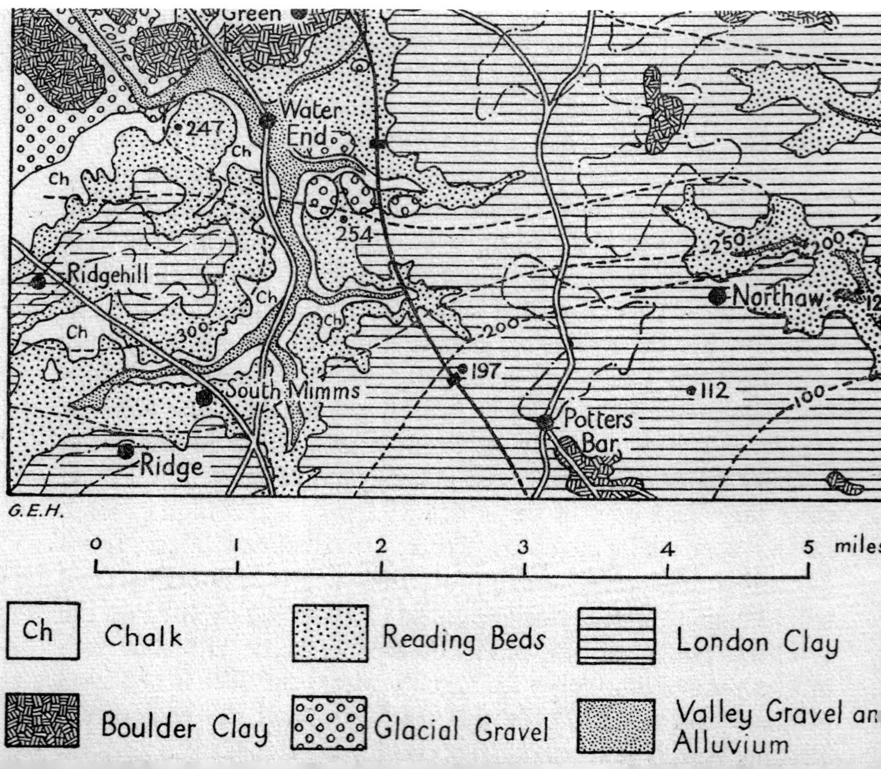

Green
R Colne
Water
End
247
Ch
254
Ridgehill
300
200
197
South Mimms
Ridge
Potters
Bar
112
100
Northaw
250
G.E.H.
0 1 2 3 4 5 miles
Ch Chalk
Reading Beds
London Clay
Boulder Clay
Glacial Gravel
Valley Gravel an
Alluvium

49

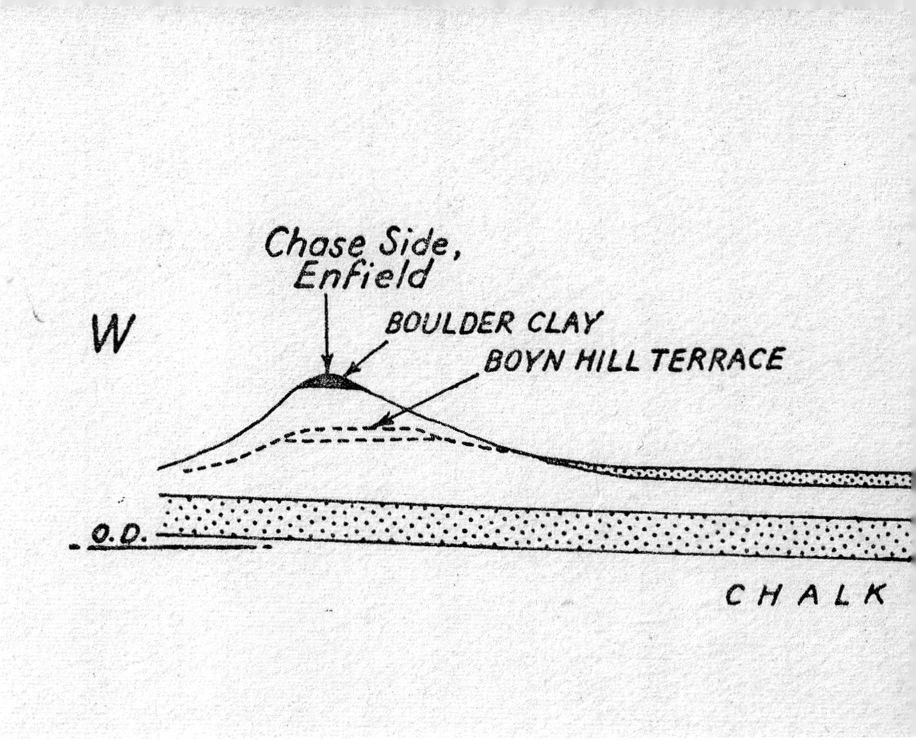
Chase Side,
Enfield
W
BOULDER CLAY
BOYN HILL TERRACE
O.D.
CHALK

50

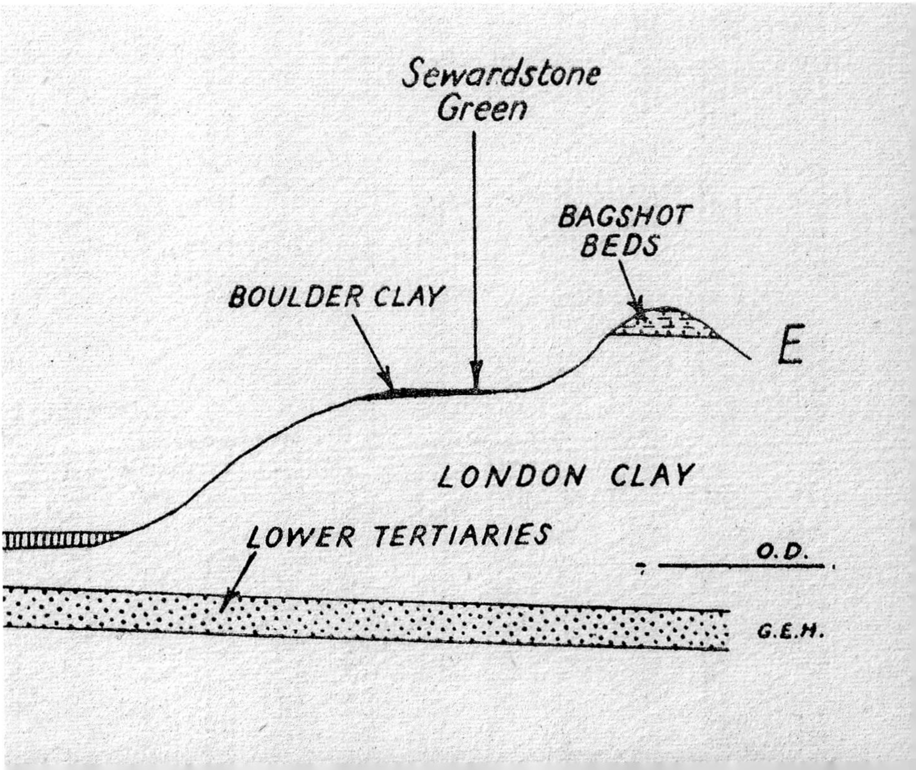
Sewardstone
Green
BAGSHOT
BEDS
BOULDER CLAY
E
LONDON CLAY
LOWER TERTIARIES
O.D.
G.E.H.

Biographical Commentary

1900
—Geoffrey Hutchings born at Strood, near Rochester in Kent, his father was a builder and amateur geologist.

1916
—Leaves Rochester Technical Junior School to begin an engineering apprenticeship at the Royal Dockyard Chatham.

1921
—Qualifies and works as an engineering draughtsman for five years.
—Joins the Rochester and District Naturalist Society. His early scientific interest was in geology and his first publication was a geological report in the society journal *The Rochester Naturalist*. At a meeting of the society he becomes acquainted with C C Fagg, an exponent of Patrick Geddes methods of regional surveying and an outstanding amateur field scientist.
—Converted to the educational and scientific objectives of regional surveying as a framework for his own field studies, coinciding with his increasing interest in teaching.

1921–1926
—Teaches science and engineering to evening classes at Medway Technical College.
—Studies at Birkbeck College, where he attends the lectures of the eminent geologist C M Davies; later he studies botany and zoology but does not proceed to a qualification.
—Meets S W Wooldridge, then Professor of Geography at Birkbeck College. Both Fagg and Wooldridge were to remain lifelong friends and collaborators of Hutchings, and they profoundly influenced his career and geographical thought.

1926
—Returns to Rochester Technical Junior School as a full time teacher, where he experiments with the teaching of biogeographical field studies at a time when science in schools was mainly confined to physics and chemistry.
—Continues to apply himself to regional surveying and acquires a detailed topographical knowledge of Kent and other parts of the south east. As Honorary Secretary of the Regional Survey Section of the South East Union of Scientific Societies, edits and writes papers describing the results of regional surveys or investigations into physical geography.

1928

—Association with the Centre for Rural Education at Hill Farm, Stockbury, in Kent. Collaborates with Christine Pugh in describing the surrounding topography in a privately published book: *Stockbury, a Regional Study in North East Kent.*

—Contributes chapters on topography to two Kent Regional Planning Schemes.

—Collaborates with C C Fagg in writing and illustrating the chapter on 'The South East' in *Great Britain, Essays in Regional Geography*, edited by A G Ogilvie. This work for many years became a standard university textbook. The other contributors were academic geographers, but Fagg and Hutchings were invited on the basis of their practical survey work.

1930

—Collaborates again with Fagg on *An Introduction to Regional Surveying*, a manual for teachers and students. This book did much to introduce geographical fieldwork into the curricula of schools and colleges as an interpretative activity rather than one of scientific research.

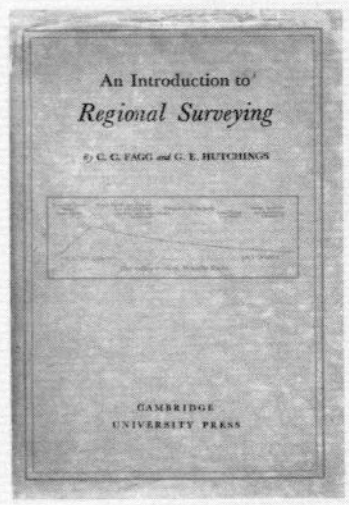

1937

—Explains the use of Hill Farm as a training place for teachers and students in regional survey in a manuscript prepared to propose its adoption as a field study centre by a local education authority or the University of London.

1938–1945

—In the Middle East serving as the Principal of Technical Schools, first in Baghdad and then Bahrain.

1943

—Writes to F H C Butler, with whom he had worked before the war on the Executive Council of the School Nature Study Union, enquiring about the possibility of establishing field study centres as part of the envisaged post-war reconstruction programme in education. Hill Farm was referred to as an example of the kind of centre that might be set up. In December, Butler calls the inaugural meeting of the Council for the Promotion of Field Studies, now the Field Studies Council. From the beginning, Hutchings is recognised as a strong candidate for the Wardenship of one of the Field Study Centres that the CPFS begins to establish.

1946

—The National Trust leases Juniper Hall, near Dorking, Surrey to the CPFS, and Fagg is made Honorary Warden until Hutchings can take up the full time post in the following year.

1947

—Appointed Warden of Juniper Hall Field Study Centre. From the beginning Hutchings identifies himself more closely with geography, and his concept of educational field studies was strongly influenced by S W Wooldridge, then Chairman of the Juniper Hall Committee and later on the CPFS Council. They both conceived of geographical field work as a method of acquiring 'an eye for country', through scientifically observing the countryside and learning to interpret the evolution of landscape. To this process Hutchings contributed his geographical landscape drawing technique derived from his early training as a draughtsman and his ability as an artist.

1949

—He describes his geographical philosophy in an address to the School Nature Study Union: 'The geographer as field naturalist', arguing that the insight of the natural historian combines with that of the physical geographer to explain the inter relationships expressed in the regional division of the earths surface and the appearance of landscape. For him the study of landscape emphasised the ideas of geography as a field science, and he was critical of what he called 'paper geographers' who culled their facts from documents only.

1952

—Writes *The Book of Box Hill*, an account of the geography, history, flora and fauna of the chalk scarp immediately south of Juniper Hall. The pamphlet is published by the 'Friends of Box Hill', an association formed in the previous year.

1955

—Publishes from Juniper Hall Field Centre the instructional booklet: *An Introduction to Geographical Landscape Drawing (Lowland Country)*. The central panoramic spread: 'Example of a landscape drawing with selected detail'—looking north east towards Box Hill and the North Downs—is a *tour de force* in the rendering of foreground and distance, and the integration of locative descriptions.

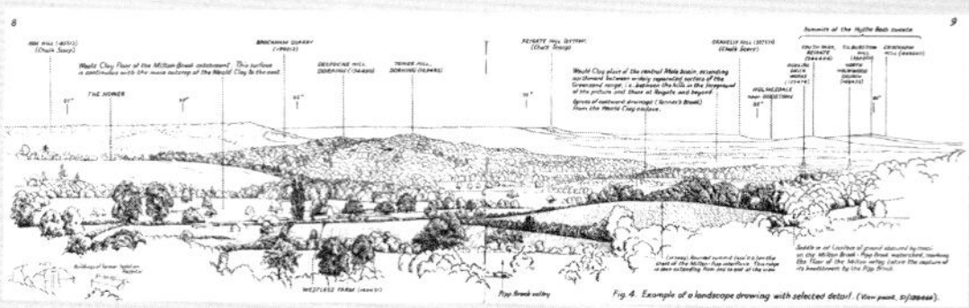

1956

—Retires from Juniper Hall partly due to ill health. Appointed CPFS Senior Tutor in Geography, enabling him to travel to the various field centres to advise the Wardens.

1957

—Collaborates with S W Wooldridge in producing *London's Countryside*, which became a model for writing up fieldwork excursions in rural areas and was subsequently adopted by numerous geographical publications.

1960

—Five years after publication of his landscape drawing booklet he expands the study into a full-scale primer, *Landscape Drawing*, with an introduction by D L Linton, Professor of Geography at the University of Birmingham and also an exponent of field sketching. Combining the exactitude of observational science with the interpretative power of an artist, the chapters compare photography and drawing; the process of setting out drawings; light and shade; perspective; panoramas; the rendering of mountains, trees and buildings; and guidance on materials. Perhaps more than any other publication arising from the work of the CPFS, it summed up the twin aims of the Council in its efforts to encourage field studies in the widest sense having both scientific and aesthetic objectives. In particular it typified Hutchings' own interdisciplinary interests, which were those of a gifted amateur who could express his scientific observations through a combination of drawing and precise prose.

1964

—Dies after a short illness. His account of a field trip by S W Wooldridge, who had died the previous year, appeared posthumously in *Guide to London Excursions*, published by the 20th International Geographical Congress: "...[Hutchings'] contribution to the teaching of geography from field observation was as distinctive and remarkable as that of S W Wooldridge. This account, with Hutchings' own maps, stands as a memorial to them both."

Much of the detail was obtained from the article on Hutchings by Keith Wheeler in *Geographers Biobibliographical Studies*, Volume 2 (London, 1978).

First published 2006
ISBN–10: 0 9537048 5 8 | ISBN–13: 978 0 9537048 5 9
Colin Sackett, 7 Hillhead Terrace, Axminster, Devon EX13 5JL
books@colinsackett.demon.co.uk
Printed by Axminster Printing Company